Computational Biology

The Computational Biology series publishes the very latest, high-quality research devoted to specific issues in computer-assisted analysis of biological data. The main emphasis is on current scientific developments and innovative techniques in computational biology (bioinformatics), bringing to light methods from mathematics, statistics and computer science that directly address biological problems currently under investigation.
The series offers publications that present the state-of-the-art regarding the problems in question; show computational biology/bioinformatics methods at work; and finally discuss anticipated demands regarding developments in future methodology. Titles can range from focused monographs, to undergraduate and graduate textbooks, and professional text/reference works.

Author guidelines: springer.com > Authors > Author Guidelines

Also in this series

Burger, A.; Davidson, D., Baldock, R. (Eds)
Anatomy Ontologies for Bioinformatics
Principles and Practice
ISBN 978-1-84628-884-5 2008 (Hardcover)

Dubitzky, Werner; Azuaje, Francisco (Eds)
Artificial Intelligence Methods and Tools for Systems Biology
Vol. 5, ISBN 978-1-4020-2859-5, 2004
(Hardcover)

Dubitzky, Werner; Azuaje, Francisco (Eds)
Artificial Intelligence Methods and Tools for Systems Biology
Vol. 5, ISBN 978-1-4020-2959-2, 2005
(Softcover)

Bininda-Emonds, Olaf R.P. (Ed.)
Phylogenetic supertrees
Combining information to reveal the Tree of Life
Vol. 4 ISBN 978-1-4020-2328-6, 2004 (Hardcover)

Bininda-Emonds, Olaf R.P. (Ed.)
Phylogenetic supertrees
Combining information to reveal the Tree of Life
Vol.4, ISBN 978-1-4020-2329-3, 2004 (Softcover)

Ramsden, Jeremy J.
Bioinformatics
An Introduction
Vol. 3, ISBN 9781-4020-2141-1, 2004
(Hardcover)

Koski, T.
Hidden Markov Models for Bioinformatics
Vol. 2, ISBN 978-1-4020-0135-2, 2002
(Hardcover)

Koski, T.
Hidden Markov Models for Bioinformatics
Vol. 2, ISBN 978-1-4020-0136-9, 2002
(Softcover)

Sankoff, D.; Nadeau, J.H. (Eds)
Comparative Genomics
Empirical and Analytical Approaches to Gene Order Dynamics, Map Alignment and the Evolution of Gene Families
Vol. 1, ISBN 978-0-7923-6584-6, 2000
(Softcover)

Sankoff, D.; Nadeau, J.H. (Eds)
Comparative Genomics
Empirical and Analytical Approaches to Gene Order Dynamics, Map Alignment and the Evolution of Gene Families
Vol. 1, ISBN 978-0-7923-6583-9, 2000
(Hardcover)

Anna Panchenko · Teresa Przytycka
Editors

Protein-protein Interactions and Networks

Identification, Computer Analysis, and Prediction

Editors
Anna Panchenko
National Institutes of Health
Bethesda, Maryland
USA
panch@ncbi.nlm.nih.gov

Teresa Przytycka
National Institutes of Health
Bethesda, Maryland
USA
przytyck@ncbi.nlm.nih.gov

Computational Biology Series ISSN 1568-2684
ISBN: 978-1-84996-731-0 e-ISBN: 978-1-84800-125-1
DOI: 10.1007/978-1-84800-125-1

British Library Cataloguing in Publication Data
A catalogue record for this book is available from the British Library

Printed on acid-free paper

Springer Science+Business Media

springer.com

Preface

In modern physics, all phenomena in the universe are considered to be the results of interactions between particles. In biology, we are now at the point where the latest advances of experimental and computational techniques have opened a new perspective. Indeed, it has been shown that living organisms are extremely complex and the numerous biomolecules work together in a coordinated fashion to provide specific cellular functions. To analyze this added level of complexity, the field of Systems Biology has emerged-the area of research that focuses on understanding the roles of interactions between genes, proteins, and other cell components.

Biological interactions, in particular protein-protein interactions, are astonishing in their magnitude and diversity. It has been discovered that the vast majority of proteins interact with multiple partners (on average with six to eight other proteins) and thousands of different proteins form intricate interaction networks or highly regulated pathways. Thanks to the abundance of high throughput experimental data, researchers have begun to uncover general rules obeyed by protein-protein interaction networks, principles of their evolution, and the means of their functioning. Analysis of patterns and principles governing protein-protein interactions prompted, in turn, a rapid development of computational methods to predict missing elements of protein interaction networks and to identify the roles of individual components of these networks in cell function.

The study of protein-protein interactions is a multidisciplinary endeavor. By putting pieces of the interaction puzzle together researchers are now able to construct genome-wide protein interaction networks, obtain insights into the physico-chemical principles of protein binding and in some cases predict protein interaction partners. All of these achievements would not be possible without an alliance between the fields of biology, physics, and computer science, which allows to look at the interaction scenarios from different angles.

This book is a collection of nine reviews written by experts from diverse scientific backgrounds, each offering a unique perspective on this rapidly developing field. It describes the most important problems in the area of protein-protein interactions and presents a spectrum of approaches to address these problems. The first chapter focuses on the experimental techniques to discover protein-protein interactions. It allows the reader to appreciate the interplay between various experimental techniques, their strengths and limitations and possible biases that may be inherent for a particular method. Much of the explosion of scientific results in this emerging field is without a doubt attributable to community-wide sharing of data and results through publicly available databases. Consequently, the second chapter of this book is devoted to protein interaction databases and methods of integrating data from diverse sources. The authors of Chapter 3, focus their attention on general principles of protein binding and common properties of interaction interfaces inferred from protein crystal structures. The next two chapters explore the methods of prediction of protein-protein and domain-domain interactions respectively, while Chapter 6 presents an integrative approach that has lead to successful reconstructions of large macromolecular complexes. The topological properties of protein interaction networks are reviewed in Chapter 7 which also describes the studies of dynamical responses of networks to perturbations. In addition, the topology of interaction networks can be used to uncover the function of uncharacterized proteins, which is the topic of Chapter 8. Finally, an extremely important source of our knowledge about living organisms comes from comparative studies. Thus, the concluding chapter of this book is devoted to cross-species comparison of protein-protein interaction networks.

It was our intention to present to the reader a book that will give an in-depth overview on the subject of protein-protein interactions. However, the reader should keep in mind that the understanding of protein interaction networks in general and the role of many individual components of these networks in particular, is far from being complete. Researchers have only begun to decipher protein-protein interaction networks and we are presenting the first snapshots that emerge from these studies. The work on this book has been extremely rewarding and we would like to thank all the contributors for making it possible.

Bethesda, Maryland

Anna Panchenko
Teresa Przytycka

Contents

Contents

Contributors

Frank Alber
Molecular and Computational Biology Program, Department of Biological Sciences, University of Southern California, Los Angeles, CA 90089-1340, USA, alber@usc.edu

Gerard Cagney
Conway Institute, University College Dublin, Belfield, Dublin 4, Ireland, gerard.cagney@ucd.ie

Brian T. Chait
Laboratory of Mass Spectrometry and Gaseous Ion Chemistry, The Rockefeller University, 1230 York Avenue, New York, NY 10021-6399, USA, chait@rockefeller.edu

Katia S. Guimarães
Center of Informatics, Federal University of Pernambuco, Recife, Brazil, katiaguim@gmail.com

Attila Gursoy
Koc University, Center for Computational Biology and Bioinformatics, and College of Engineering, Rumelifeneri Yolu, 34450 Sariyer Istanbul, Turkey, agursoy@ku.edu.tr

Ozlem Keskin
Koc University, Center for Computational Biology and Bioinformatics, and College of Engineering, Rumelifeneri Yolu, 34450 Sariyer Istanbul, Turkey, okeskin@ku.edu.tr

Sergei Maslov
Brookhaven National Laboratory, Department of Condensed Matter Physics and Materials Science, Upton, New York, USA, maslov@bnl.gov

Ruth Nussinov
Basic Research Program, SAIC-Frederick, Inc. Center for Cancer Research Nanobiology Program, NCI-Frederick, Frederick, MD 21702; Sackler Institute of Molecular Medicine, Department of Human Genetics and Molecular Medicine,

Sackler School of Medicine, Tel Aviv University, Tel Aviv 69978, Israel, ruthn@ncifcrf.gov

Anna Panchenko
National Center for Biotechnology Information, National Institutes of Health, Bethesda, USA, panch@ncbi.nlm.nih.gov

Florencio Pazos
Computational Systems Biology Group, National Center for Biotechnology (CNB-CSIC), Madrid, Spain, pazos@cnb.uam.es

Teresa M. Przytycka
National Center for Biotechnology Information, National Institues of Health, Bethesda, USA, przytyck@mail.nih.gov

Michael P. Rout
Laboratory of Cellular and Structural Biology, The Rockefeller University, 1230 York Avenue, New York, NY 10021-6399, USA, rout@mail.rockfeller.edu

Eytan Ruppin
School of Computer Science and School of Medicine, Tel-Aviv University, Tel-Aviv 69978, Israel, ruppin@post.tau.ac.il

Andrej Sali
Departments of Biopharmaceutical Sciences and Pharmaceutical Chemistry, and California Institute for Quantitative Biosciences, Byers Hall, Suite 503B, University of California at San Francisco, 1700 4th Street, San Francisco, CA 94158-2330, USA, sali@salilab@org

Roded Sharan
School of Computer Science, Tel-Aviv University, Tel-Aviv 69978, Israel, roded@post.tau.ac.il.

Benjamin Shoemaker
National Center for Biotechnology Information, National Institutes of Health, Bethesda, USA, shoemake@mail.nih.gov

Mona Singh
Department of Computer Science and Lewis-Sigler Institute for Integrative Genomics, Princeton University 08544, USA, mona@cs.princeton.edu

Björn Titz
Forschungszentrum Karlsruhe, Box 3640, 76021 Karlsruhe, Germany, bjoern.titz@itg.fzk.de

Peter Uetz
The J Craig Venter Institute, 9712 Medical Center Drive, Rockville, MD 20850, USA, uetz@jcvi.org

Alfonso Valencia
Structural Computational Biology Programme, Spanish National Cancer Research Centre (CNIO), Madrid, Spain, valencia@cnio.es

Nir Yosef
School of Computer Science, Tel-Aviv University, Tel-Aviv 69978, Israel, niryosef@post.tau.ac.il

Chapter 1
Experimental Methods for Protein Interaction Identification and Characterization

Peter Uetz, Björn Titz, and Gerard Cagney

Abstract There are dozens of methods for the detection of protein-protein interactions but they fall into a few broad categories. Fragment complementation assays such as the yeast two-hybrid (Y2H) system are based on split proteins that are functionally reconstituted by fusions of interacting proteins. Biophysical methods include structure determination and mass spectrometric (MS) identification of proteins in complexes. Biochemical methods include methods such as far western blotting and peptide arrays. Only the Y2H and protein complex purification combined with MS have been used on a larger scale. Due to the lack of data it is still difficult to compare these methods with respect to their efficiency and error rates. Current data does not favor any particular method and thus multiple experimental approaches are necessary to maximally cover the interactome of any target cell or organism.

1.1 Introduction

Protein interactions can be identified by a multitude of experimental methods. In fact, the IntAct database of molecular interactions currently lists about 170 different experimental methods and variations thereof that can be used to detect and characterize protein-protein interactions (the main classes are listed in Table 1.1). While we present the commonly used methods in this chapter we will focus on the few technologies which are used in high-throughput studies and thus generated the vast majority of interaction data available today: the yeast two-hybrid assay and protein complex purification and identification by mass spectrometry (MS) (Table 1.2). These two methods represent two fundamentally different sources of interaction data and thus it is important to understand how they work and what strengths and weaknesses each of them has. This is

P. Uetz
The J Craig Venter Institute, 9712 Medical Center Drive, Rockville, MD 20850, USA
e-mail: uetz@jcvi.org

A. Panchenko, T. Przytycka (eds.), *Protein-protein Interactions and Networks*,
DOI: 10.1007/978-1-84800-125-1_1,

Table 1.1 Methods to detect protein-protein interactions, based on the PSI MI classification. Listed are the top categories with important examples. The whole list contains more than 170 terms and can be found at http://www.ebi.ac.uk/intact/ (go to Advanced Search > detection methods). Methods in *italics* are discussed or illustrated in this chapter

- **protein complementation assay**
  - *cytoplasmic complementation assay*
    - *ubiquitin reconstruction*
  - *membrane bound complementation assay*
    - *mammalian protein protein interaction trap*
  - *transcriptional complementation assay*
    - *two hybrid*
  - *bimolecular fluorescence complementation*
  - 3 hybrid method
    - protein tri hybrid
- **biophysical**
  - nuclear magnetic resonance
  - surface plasmon resonance
  - *mass spectrometry studies of complexes*
  - x-ray crystallography
  - isothermal titration calorimetry
  - fluorescence technology
    - fluorescent resonance energy transfer (FRET)
- **biochemical**
  - cross-linking study
  - *affinity technology*
    - display technology
    - *far western blotting*
    - *affinity chromatography technology*
      - *pull down*
      - *tandem affinity purification*
      - *coimmunoprecipitation*
    - *array technology*
      - *peptide array*
      - *protein array*
  - enzymatic study
    - phosphotransfer assay
- **imaging techniques**
  - fluorescence microscopy

especially important for theoretical analyses which often draw conclusions from datasets which may not be adequate for certain studies. For example, membrane proteins are underrepresented in both yeast two-hybrid and complex purification studies.

Table 1.2 The contribution of various PPI methods to protein interactions in the IntAct database (as of Sep 8, 2007)

Method	number of interactions	Percent
Two-hybrid	62,340	63.4%
Co-IP	8220	8.4%
TAP purification	4475	4.6%
Other	23250	23.6%
Total	98285	100%

1.1.1 Complex Versus Binary Interactions

It is important to note that most methods detect either direct binary interactions or indirect interactions without knowing which proteins are interacting. The yeast two-hybrid system usually detects direct binary interactions while complex purification detects the components of complexes (Fig. 1.1). Complex data are often interpreted as if the proteins that co-purifiy are interacting in a particular manner, consistent with either a spoke or matrix model. The **spoke model** assumes that all proteins in a complex interact with the bait protein only while the **matrix model** assumes that all proteins interact with all others. Even a combination of both methods is usually not sufficient to establish the precise topology as some interactions may be too weak to be detected individually. X-ray crystallography can provide a detailed model of the proteins in a complex. However, note that crystallized complexes often lack additional weakly associated proteins that do not co-crystallize and thus may not provide a complete picture either.

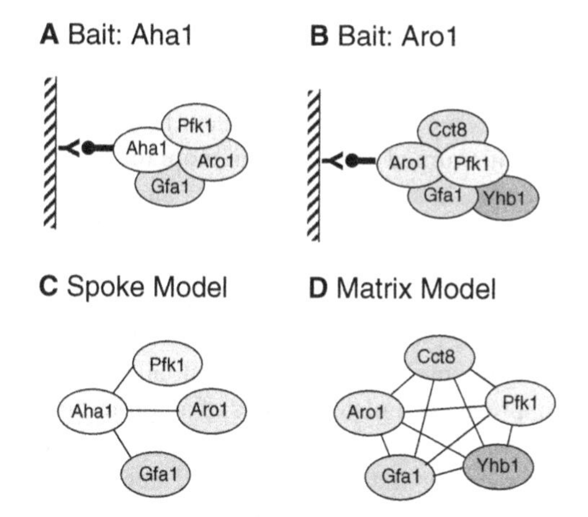

Fig. 1.1 Protein complexes vs. binary interactions. (**A,B**) When two proteins of a complex are tagged and the other components identified, the two purifications rarely result in the same components. (**C,D**) Although proteins in a complex are associated, it remains usually unclear which proteins interact directly with each other. In order to predict direct interactions either the matrix (**C**) or spoke model (**D**) is applied to lists of purified proteins. To evaluate such interactions Gavin et. al have invented the socio-affinity index (SAI). In brief, the SAI quantifies the tendency for a protein pair (e.g. Aro1 and Gfa1) to identify each other when tagged (**B**) and to co-purify when other proteins are tagged (**A**) relative to what would be expected from their frequency in the data set. High affinity values result when both proteins co-purifiy when either one of them is tagged (without purifying many other proteins) and when both are always seen together in purifications of other baits. Modified after (Goll and Uetz 2006)

1.1.2 The Biological Relevance of Detected Protein-protein Interactions

There is considerable debate on the relevance of experimentally detected interactions. Given the fact that only about 20% of all yeast proteins and even smaller percentages of all proteins in bacteria are essential under laboratory conditions, it is clear that an even smaller number of all detectable interactions are essential for survival. Thus, a large fraction of all detected protein-protein interactions may only be required under specific biological conditions. Alternatively, they may not be relevant to a cell at all, for instance when two proteins that interact in vitro never interact in vivo because they are housed in different cellular compartments. Indeed, a major challenge for the future will be to distinguish "essential" from "non-essential" interactions and then to identify the non-essential interactions that have a biological role and thus provide a selective advantage. It is possible that a class of non-essential and "irrelevant" interactions are continuously generated and lost in the course of evolution but only occasionally selected. As long as they do not harm the cell they are simply subject to loss through random genetic drift.

1.1.3 Protein-protein Interactions are Incompletely Studied

A complete description of protein-protein interactions would require the structure of the proteins involved. Because proteins come together to carry out biochemical functions, ideally we would also know their localization, precise concentration, and how the genes of their components are regulated, how stable the proteins are and thus how quickly they are turned around. Even more importantly, we would need to know the precise affinities and thus the dynamics and kinetics of complex assembly. Assembly of complexes often involve conformational changes about which we know very little. Neither do we fully understand the role of post-translational modifications and how they affect the assembly of protein complexes. We should keep in mind that we are still in the process of qualitatively cataloging protein-protein interactions without paying too much attention to quantitative and dynamic aspects. This will change as we approach complete catalogs of all protein-protein interactions for the major model systems. Some recent studies estimate that we have identified only 50% of all yeast interactions and only 10% of all human interactions (Hart et al. 2006). We cannot make such estimates for other species for which there is still too little information.

1.2 Protein Complementation Techniques

The most popular protein complementation technique is the yeast two-hybrid system. All such complementation techniques are based on the reconstitution of split proteins that re-generate a functional protein from two halves. After the yeast two-hybrid system was invented, researchers realized that they can apply its concept to many other proteins. In fact, new complementation techniques continue to be invented. This chapter will focus on the classical yeast two-hybrid method as it is the only one that has been applied to a large number of protein interactions while the utility of the other methods is still being investigated.

1.2.1 The Yeast-Two-Hybrid System

The yeast-two-hybrid (Y2H) system is a widely used genetic assay for the detection of protein-protein interactions. The original assay was developed by Fields and Song (1989) and takes place in living yeast cells (Fig. 1.2A). It employs a transcription factor, e.g. the yeast transcription factor GAL4, which can activate a reporter gene when its DNA-binding domain (DBD) and its transcriptional activation domain

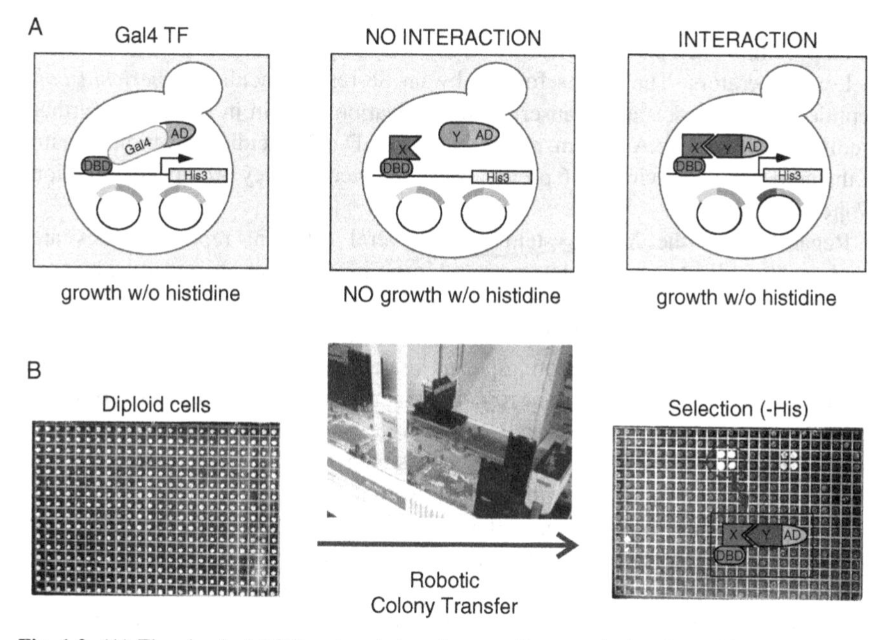

Fig. 1.2 (**A**) The classical Y2H system is based on a split transcription factor (Gal4 TF). In the native Gal4-TF a DNA-binding domain (DBD) is covalently bound to an activation domain (AD). The transcription factor activates the expression of a reporter gene (his3) in specially engineered yeast cells, which allows for growth under histidine-deficient conditions. For the Y2H assay, a protein X is fused to the DBD and a protein Y to the AD of Gal4. If X-Y do not interact, no growth without histidine is possible. However, when X binds to Y, an active transcription factor is reconstituted and the respective yeast cells can grow on histidine-deficient medium. (**B**) For the array-based Y2H system individual yeast colonies are arrayed onto agar plates in the 384-well format. Protein-pairs, which are to be tested for an interaction, are combined at each position of the yeast array by a mating based approach (in diploid yeast cells). This yeast array is transferred to selective conditions (e.g., w/o histidine) employing a robotic procedure. Only at positions of the array, which carry an interacting protein-pair, yeast colonies can grow (Note that the shown test was done in quadruplicates)

(AD) are linked. When both domains are separated from each other, they do not have the capability to activate transcription of the reporter gene. To answer the question whether a protein A interacts with protein B, each protein is fused to one of these transcription factor domains, AD and DBD, respectively. If protein A binds to protein B, an active transcription activator complex is re-established, the reporter gene is transcribed and its gene product can be used to detect the protein-protein interaction. The protein linked to the DNA-binding domain is called bait; the protein linked to the activation domain is called prey.

In the original Y2H system developed by Fields and Song (1989) both fusion constructs are derived from the yeast GAL4 protein: baits contain the DNA binding domain (amino acids [aa] 1–147) and preys the activation domain (aa 768–881) of this transcriptional regulator.

However, any other transcription factor can be used as well. The lexA-based system is just one alternative that was developed by the laboratory of Roger Brent (Golemis and Khazak 1997). In this system, the DBD is provided by the prokaryotic LexA protein, which physiologically acts as a transcriptional repressor, when bound to LexA operators. The AD is formed by an 88-residue acidic *Escherichia coli* peptide (B42) that acts as a transcriptional activation domain in yeast. In a further modification of the lexA system, either the Gal4-AD or an acidic activation domain of the herpes simplex virus V16 protein was employed for prey protein construction (Vojtek et al. 1993).

Regardless of the Y2H system used, several different reporter genes are employed to indicate a protein interaction. Most reporter genes are under the control of artificial promoter constructs, which consist of an appropriate UAS (upstream activation sequence) and a TATA sequence. Most reporter genes are integrated into the yeast genome. Therefore, special yeast strains are used for Y2H tests. β-galactosidase (lacZ) is a popular reporter, since several different assays for measuring its activity are available and quantitative values of the reporter activation can be obtained. However, lacZ is not well suited for library screening. Rather, reporters are chosen that allow the cell to grow if it harbors interacting bait and prey proteins. These reporters are often metabolic enzymes which are deleted or mutated in the Y2H yeast strain. Common reporter enzymes are involved in histidine metabolism (His3), leucine metabolism (Leu2), adenine metabolism, and uracil metabolism (Ura3). This way, an interaction can be simply selected by growing the Y2H strain on media lacking histidine or leucine. Different Y2H systems use different sets of reporters. Their promoters vary slightly and testing for activation of different reporter genes is thought to reduce the number of false positives.

The Y2H system can be used for testing individual protein interactions. However, its most powerful application is the screening for protein interactions in large or genome-wide libraries. Traditionally, cDNA or genomic DNA prey libraries – the latter for bacteria – are constructed. These mixed prey libraries are tested against a specific bait construct – by either a co-transformation or a mating based strategy. Under selective conditions, only yeast strains carrying an interacting bait/prey pair can grow. DNA sequencing is required for the identification of the prey library clone.

An alternative approach is the array-based Y2H system (Cagney et al. 2000). The first step of this system is the systematic cloning of all open-reading frames of a genome (or a subset thereof) into prey vectors. These vectors are transformed into haploid yeast cells individually and the resulting prey strains are systematically arranged on culture plates with 96 or 384 positions. A haploid bait strain is then mated with all the prey strains at each position of this array; the positions of the resulting diploid cells that express both bait and prey fusions are retained throughout the assay and individual protein-interactions can easily be identified by their reporter-gene activation at a specific positions of the array (Fig. 1.2B).

Mixed library screens do not require systematic cloning of all prey constructs, however, the prey library must be created. Therefore, the complete DNA sequence of the genome of interest is no prerequisite. Whereas the setup of the prey library is more time-consuming in the array-based Y2H system, the final workup of the interactions is faster: the position of the growing yeast colony on the array directly identifies the interacting protein pair. In addition, the array-based system can be controlled much better. Every bait is tested against the same set of prey proteins – on the contrary, the bait-prey combinations in the random library screen are defined by a random process. Thus, the signal-to-background ratio can be systematically evaluated in an array screen for each bait protein and the specificity of each interaction can be known, e.g. the number of proteins a given prey is interacting with (Fig. 1.2B). This reduces the number of experimental false-positives (see below).

Pooled-library screens combine both presented approaches. In this strategy, preys of known identity (systematically cloned or sequenced cDNA library clones) are combined and tested as pools against bait strains. The identification of the interacting protein pair commonly requires either sequencing or retesting of all members of the respective pool. Zhong et al. established a method, which allows for pooling up to 96 preys (Zhong et al. 2003). It was estimated that this pooling scheme reduces the number of interaction tests required to 1/8–1/24 in the case of the yeast proteome. Two recent large-scale interaction mapping approaches for human proteins employed such a pooling strategy: Rual et al. tested baits against pools of 188 preys and identified individual interactions by sequencing (Rual et al. 2005); Stelzl et al. tested pools of 8 baits against a systematic library of individual preys and identified interactions by a 2nd interaction mating (Stelzl et al. 2005). Recently, an "smart-pool-array" system was proposed, which allows the deconvolution of the interacting pairs through the definition of overlapping bait pools (Jin et al. 2007), and thus usually does not depend on sequencing or a 2nd pair-wise mating procedure.

1.2.2 Other Fragment Complementation Techniques

Several other methods for the detection of protein interactions rely on the co-expression of two-hybrid fusion proteins (Fig. 1.3, Table 1.3). All these methods have been proven to work with a selected set of protein interactions. Unfortunately, no systematic attempts have been undertaken to compare the quality and methodological biases of these approaches. Some approaches offer additional advantages

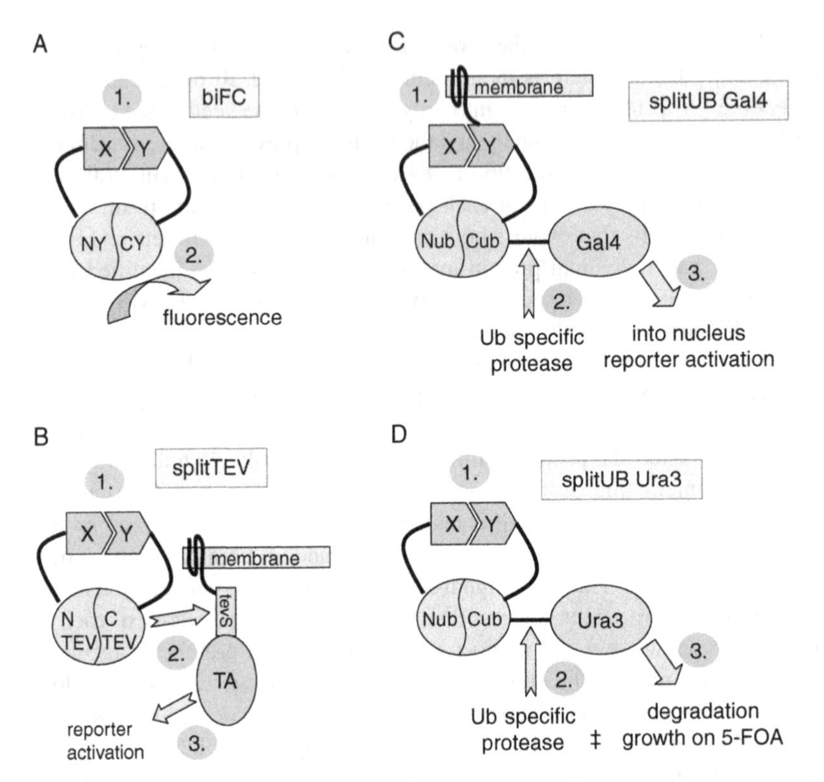

Fig. 1.3 Fragment complementation assays. (**A**) In bimolecular fluorescence complementation (biFC) methods the interaction of X and Y leads to the reunion of two non-fluorescent-protein fragments into a functional fluorophore. (**B**) In the splitTEV method the interaction results in the formation of an active TEV protease, which can, for example, release a membrane bound transcription factor. (**C**) In split-ubiquitin methods reunited ubiquitin is recognized by ubiquitin-specific proteases. This can lead to release of membrane bound transcription factors or (**D**) to the degradation of an enzyme (Ura3), which mediates toxicity of 5-FOA

Table 1.3 Variations of the yeast two-hybrid system

Class	Method	Principle	Reference
Y2H	classical Y2H	Reconstitution of active transcription factor, here based on Gal4 transcriptional regulator	(Fields and Song 1989)
Y2H	lexA-based Y2H	Reconstitution of active transcription factor, based on lexA (DBD) and VP16 or Gal4 (AD)	(Vojtek et al. 1993; Golemis and Khazak 1997)
Y2H	SOS recruitment system	Activation of Ras signaling pathway made dependent on interaction	(Aronheim 1997)
B2H	split adenylate cyclase	Reconstitution of adenylate cyclase	(Karimova et al. 1998)

Table 1.3 (continued)

Class	Method	Principle	Reference
B2H	RNA Polymerase recruitment	Activation of reporter gene by RNA polymerase recruitment (similar to Y2H)	(Joung et al. 2000)
M2H	MAPPIT	Activation of cytokine signaling	(Eyckerman et al. 2001)
M2H	mammalian two-hybrid system	Reconstitution of active transcription factor	(Luo et al. 1997)
FC	split-ubiquitin (splitUB)	Protein fragment complementation: analysis of membrane proteins	(Johnsson and Varshavsky 1994; Stagljar et al. 1998)
FC	split TEV protease	Protein fragment complementation: flexible choice of reporter system	(Wehr et al. 2006)
FC	biFC	Protein fragment complementation: fluorescent proteins (allows to localize an interaction)	(Hu and Kerppola 2003)
3H	Three hybrid/kinase co-expression	Classical Y2H with kinase co-expression (detects phosphorylation dependent interactions)	(Marti et al. 1998)

Y2H = Yeast two-hybrid, B2H = Bacterial two-hybrid, M2H = mammalian two-hybrid. FC = fragment complementation, 3H = three-hybrid.

such as the localization of protein-interactions as in the case of "bimolecular fluorescence complementation" but also require additional equipment such as fluorescent microscopes; other differences are subtle and it remains to be seen how they compare in high-throughput screens. For the lack of comparative data we do not discuss these methods here. Readers are referred to the literature cited in Table 1.3 for more details.

1.3 Affinity Purification Methods

While protein complementation techniques are usually used in vivo, affinity purification requires that the interacting proteins be purified from a cell and then identified in vitro (even though the interaction takes place in vivo). Historically, GST pulldowns (see below) and co-immunoprecipitation (co-IP) have been the most popular methods, although they have been supplanted by refined high-throughput methods that use mass specrometry for protein identification. However, all these methods are based on the principle that interactions involving affinity-tagged proteins formed in vivo are preserved during biochemical purification steps. Thus we introduce GST pulldowns and co-IPs first.

1.3.1 GST-Pulldown

A standard method for in vitro interaction assays uses Glutathion-S-Transferase (GST) as a tag (Fig. 1.4). Traditionally GST pull-downs have been used to verify interactions that were found in two-hybrid screens and other screening procedures. For re-testing purposes, the two proteins are expressed in a heterologous system, e.g. human proteins in *E. coli*, so that additional interacting proteins are not co-purified. While it is often desirable to co-purify all members of a complex (see below), in this case we want to have only two defined proteins present in the experiment.

GST fusion proteins can be easily expressed and purified from *E. coli* by running a cell extract through a matrix of glutathione-coated beads, usually glutathione sepharose. Only GST fusion proteins and a few cellular glutatione-binding proteins bind to this matrix. Non-specifically bound proteins can be washed off with a salt solution such as PBS. Usually the fusion protein can be left on the matrix and incubated in a second protein solution, either a purified protein or an extract. Proteins from this solution will bind to the GST fusion protein. Often radio-labelled proteins are used (which can be generated by in vitro transcription/translation from a PCR fragment containing a promoter and the ORF of the protein in question). Commercial kits are available for such in vitro translation reactions to which only the PCR product and radiolabelled methionine has to be added. Alternatively epitope-tagged proteins can be used that can be detected by Western blotting.

In either case, the tagged or labelled protein is mixed with the matrix-bound GST fusion proteins and incubated. Subsequently the beads are washed so that only the GST fusion protein and the bound interacting protein are retained. Note that the concentration of salt in the washing buffer influences the experiment because it determines the stringency through progressive disruption of electrostatic interactions as the salt concentration increasese. The next step of the experiment is to boil the glutathione sepharose in sample buffer (containing SDS = sodium dodecyl sulphate as a detergent) and to separate the protein solution on a poly-acrylamide gel (SDS-PAGE = SDS polyacrylamide gel electrophoresis). If the sample contains enough protein it can be stained (e.g. with Coomassie Blue dye) in the gel and thus its molecular mass determined. However, often the amount of protein is not sufficient for staining. In such cases the protein needs to be blotted onto a membrane and detected by Western blotting or by mass spectrometry.

1.3.2 Co-Immunoprecipitation

Co-immunoprecipitation ("co-IP") is very similar to GST pull-downs (Fig. 1.5). However, instead of glutathione sepharose co-IPs usually use a sepharose matrix coated with protein A, a protein originally isolated from *Staphylococcus aureus*. Protein A binds with high affinity to the constant chains of IgG antibodies and thus sepharose-protein A columns can be easily coated with antibodies of any specificity. Such a matrix can now be incubated with proteins, e.g. from a cell or organ extract. All proteins from this extract that are recognized by the antibody bind to the matrix.

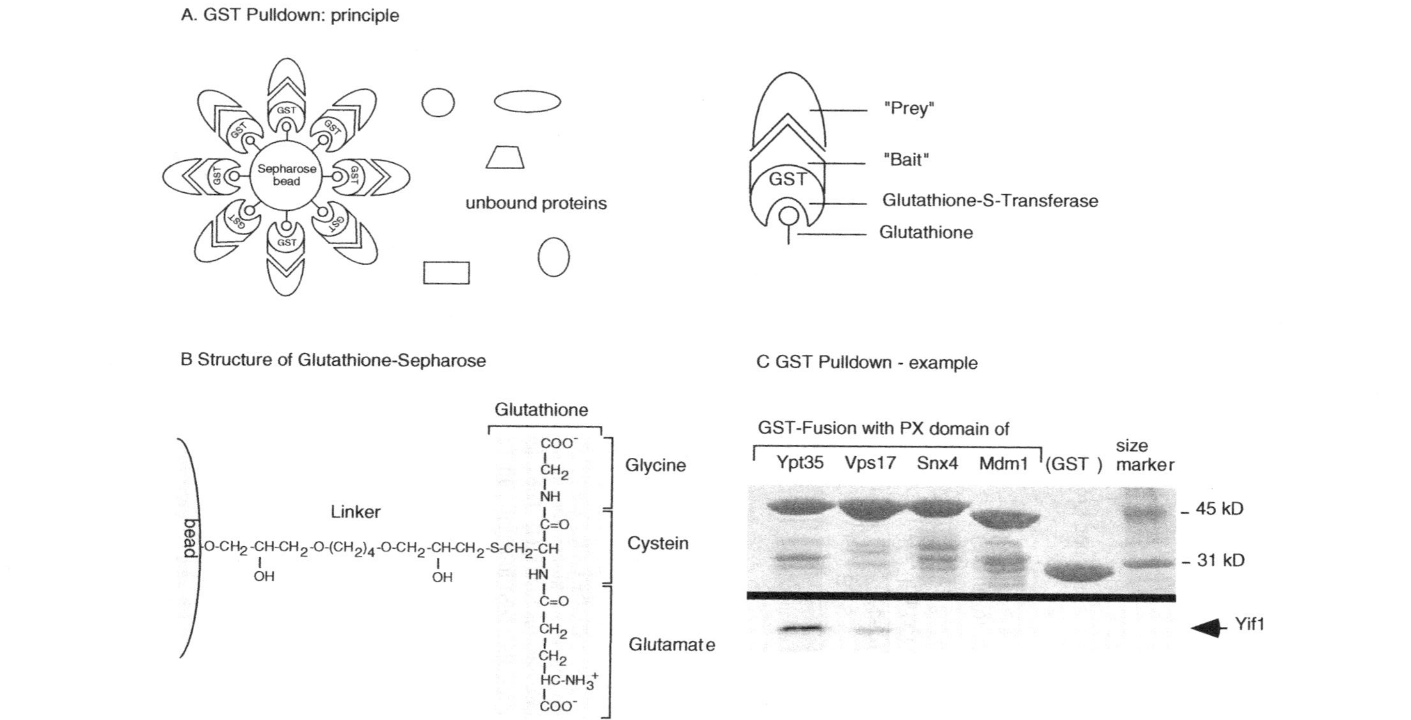

Fig. 1.4 GST-Pulldown. (**A**) General principle (see text for details). (**B**) Structure of Glutathione-Sepharose. Note that Glutathione is a natural tripeptide, which is also stable as oxidated dimer when 2 molecules are react through their SH groups. (**C**) Example: the yeast protein Yif1 binds to several PX domains of yeast. Four PX domain-GST fusion proteins as well as GST are visible in the top part of the gel (Coomassie stained). The lower panel shows an autoradiogram of radiolabelled Yif1 protein that binds to the PX domains of Ypt35 and Vps17. Yif1 was translated and radiolabelled in vitro and then incubated with bead-coupled GST fusion protein, washed, separated on a gel and then exposed to X-ray film. Modified from (Vollert and Uetz 2004)

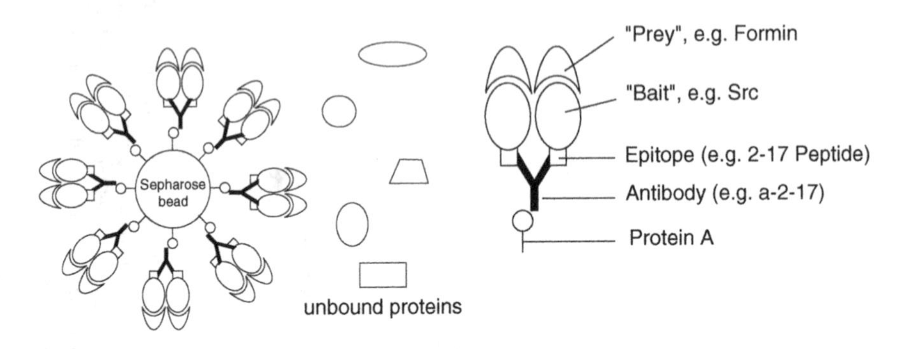

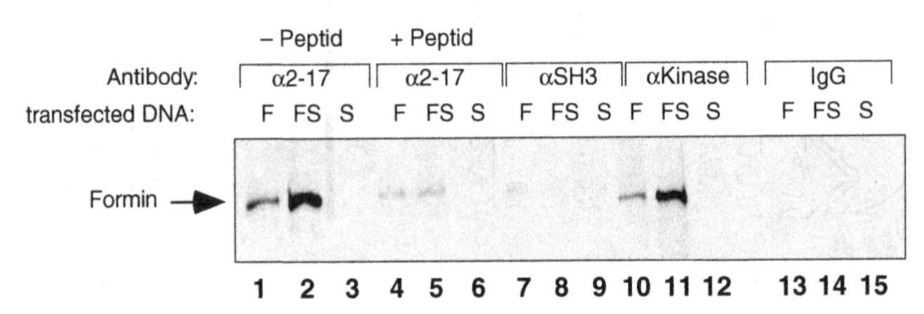

Fig. 1.5 Co-immunoprecipitation. (**A**) General principle (see text for details). (**B**) Example: A protein (here: formin) was co-precipitated with the oncoprotein Src. In this experiment four different anti-Src-antibodies have been used: one against Src peptides 2-17 (α-2-17), one against the SH3 domain (α-SH3), one against the kinase domain (α-kinase) and one control antibody mixture without binding specificity towards Src (IgG). The interacting proteins Formin and Src were expressed separately (F, S) or together (FS) in tissue culture. Cells were then lysed and incubated with the bead-bound antibodies. After washing the beads and elution in sample buffer the antibody-bound proteins were separated on a gel, Western-blotted, and detected using an anti-Formin antibody. Note that Formin cannot be co-precipitated with the anti-SH3-antibody because this antibody competes with Formin for a binding site on the SH3 domain. The peptide 2-17 also competes for the same binding site as addition of peptides (+peptide) can block binding of Formin to Src. Modified after (Uetz et al. 1996)

All other proteins can be removed by washing with buffer. Bound proteins can now be detected by boiling the matrix in sample buffer and subsequent separation on a protein gel and Western blotting. Alternatively, proteins can be identified by mass spectrometry.

The requirement for specific antibodies is currently a major limitation of co-IPs although this may be relieved in the future by commercial production of antibodies against all proteins of a genome. In addition, new technologies have emerged that are not dependent on antibodies but rather use other proteins that can be engineered

Table 1.4 Commonly used peptide affinity tags (see text for details)

Affinity tag	Capture reagent	Sequence
FLAG	Monoclonal antibody	DYKDDDDK
c-myc	Monoclonal antibody	EQKLISEEDL
S-tag	S-fragment of RNaseA	KETAAAKFERQHMDS
Strep II	Streptavidin variant	WSHPQFEK
poly-His	Ni2+-NTA	HHHHHHHH
poly-Arg	Cation exchange media	RRRRR
Calmodulin-binding domain	Calmodulin	KRRWKKNFIAVSAANRFKKISSSGAL

to have binding specificity for almost any given protein or small molecule (e.g. "affibodies" which are based on genetically engineered protein A).

Co-IPs are often used to confirm yeast two-hybrid interactions. If antibodies are not available, proteins can be labelled by specific epitopes such as hemagglutinin (HA) or myc peptides for which commercial antibodies are available (see Table 1.4). In fact, all yeast proteins have been epitope-tagged, purified and their interacting proteins identified by mass spectrometry.

1.4 Protein Complex Purification and Mass Spectrometry

The GST-pulldown and co-immunoprecipitation approaches have been improved using novel affinity tags and automated procedures for protein identification. These approaches are treated separately here but in biochemical terms are similar in principle to the pull-down and co-IP protocols described above.

1.4.1 Purification of Proteins Using Affinity Tags

Purification of proteins can be carried out under conditions that preserve stable interactions with accompanying proteins. The proteins can later be identified using methods like western blot or mass spectrometry (see below). The standard approach is to use an antibody that recognizes the protein of interest ("bait" protein) immobilized on solid phase media (e.g. sepharose beads) that are packed into a chromatography column, through which a cell lysate or protein mixture is passed. Alternatively, the media is suspended in the cell lysate. The bait is allowed to bind accompanying proteins for a period of time sufficient for equilibrium to be established, after which non-bound proteins are washed away. The washes may vary in stringency, and are sometimes applied in steps of increasing stringency and at other times as a continuous gradient. The eluted proteins are recovered and identified using approaches described below. For high-throughput projects involving dozens or hundreds of baits, more generic approaches are needed, that are independent of the requirement for production of an individual antibody or binding reagent for each bait. A wide range of affinity tags have been developed for this purpose.

Affinity tags are genetically-encoded protein fragments that can readily be recovered and are expressed as a fusion with the bait protein. Desirable properties of an ideal tag are:

- compact (so that it does not disrupt the functions of the bait or its interactions)
- high affinity for a capture reagent, so that is can ideally be recovered in a single step
- compatible with economic recovery methods
- non-toxic
- couples to a capture reagent that is non-reactive with endogenous cellular proteins.
- can be readily assayed during purification
- works for all proteins

No single affinity tag satisfies all these desired properties, and a range of strategies are used to express individual proteins, and to recover expressed proteins as well as their bound partners. Affinity tags can be broadly classified into small peptides (e.g. FLAG, poly-His) or large peptides/proteins (e.g. glutathione S-transferase, calmodulin-binding domain) (Terpe 2003). Small peptide tags are less likely to alter the tertiary structure, disrupt the function of the bait, or to be immunogenic. Larger peptides or proteins may increase the solubility of the bait but may need to be removed for applications such as antibody generation or crystallization. Some common tags are summarized in Tables 1.4 and 1.5.

The **poly-His tag** is often used for protein expression because it consists of a short tag (6–8 residues) that can be recovered using immobilized metal affinity chromatography (IMAC) systems that house nickel or other divalent metal ions. These form coordinate bonds with the histidine side chains (Porath et al. 1975), and the bait and interacting proteins can be recovered by lowering the pH by adding imidazole. The **FLAG tag** is an eight-residue hydrophobic peptide that is recognized by a number of antibodies with slightly different binding properties (M1, M2, M5). Tandem FLAG peptide units (e.g. 3 x) are often employed for increased affinity. The bait proteins are eluted by competition with a synthetic FLAG peptide or using low pH. The **c-myc tag** is an eleven residue epitope from the c-myc protein that is also bound with high specificity by an antibody (named "9E10" after its affinity). The **S-tag** technology is based on an interaction between the 15-residue S-tag and a ~100 residue S-protein fragment, both derived from RNaseA, so that assays based on the activity of this enzyme can be used to monitor the purification. The interaction

Table 1.5 Commonly used protein affinity tags (see text for details)

Affinity tag	Capture reagent
Cellulose-binding domain (CBD)	Cellulose
Chitin-binding domain (CBD)	Chitin
Glutathione S-transferase (GST)	Glutathione
Maltose-binding protein (MBD)	Amylose
Green fluorescent protein (GFP)	Monoclonal antibody

is very strong but is disrupted by strongly acid conditions. The 26-residue **calmodulin-binding peptide** (CBP) binds calmodulin in the presence of calcium, and the interaction can be disrupted using the chelator EGTA. Several types of protein are able to bind **cellulose** with high affinity, some irreversibly. Severe conditions are generally required for elution of cellulose-binding protein, involving denaturing agents, so this tag is not suitable for detecting protein interactions.

Glutathione S-transferase (GST) is widely used for protein expression and protein interaction studies (Ron and Dressler 1992). The GST protein is quite large (26 KDa) and dimerizes, but binds with high affinity to reduced glutathione. Binding is tight under non-denaturing conditions, so that bait-prey protein interactions may be maintained. The bacterial proteins **Protein A** and **Protein G** (from *Staphylococcus* and *Streptococcus* sp respectively), bind with high affinity to the Fc portions of immunoglobulins. Many other epitopes have been used effectively as affinity tags, including **V5** from bacteriophage **T7** and the **HA** tag from hemaglutinin A.

Biotinylation is often used to label biological compounds for subsequent capture due to the extremely high affinity between biotin and streptavidin ($K_a \sim 10^{-15}$M). Until recently, the introduction of the biotin group was carried out chemically, effectively precluding in vivo applications in protein interaction studies. The biotin ligase protein (BirA) from *Escherchia coli* can be used to biotinylate a lysine side-chain within a 15 residue peptide (termed "biotin acceptor peptide"). By expressing this tag as a fusion with the bait protein in a cell expressing BirA, the bait can be biotinylated in vivo, allowing effecting capture of even poorly expressed proteins from complex cell lysates.

1.4.2 Tandem Affinity Tagging

Bernard Seraphin and coworkers pioneered the use of tandem tags ("Tandem Affinity Purification", TAP), separated by proteolytically cleavable regions (Rigaut et al. 1999). After binding of the bait and associated proteins to chromatography media via one tag, the media is washed and a protease that recognizes and cleaves a sequence in the inter-tag region is introduced. This results in release of the bait which still retains the second tag (Fig. 1.6). A subsequent step introduces media with affinity for the second tag. The advantage of this tandem approach is that very stringent conditions can be used to ensure that minimal background binding (by non-specific proteins) takes place.

Potential tags can be drawn from the list discussed above, but effectively the FLAG, His, HA, Protein A, myc, and calmodulin-binding domain tags have been used in many systems because they can bind under non-denaturing conditions where interactions of the bait with associated proteins can be maintained. The initial description of the TAP method used Protein A and calmodulin-binding domain tags separated by a tobacco etch virus (TEV) protease cleavage site. The TEV protease recognizes a seven residue sequence (EXYXQS, cleavage C-terminal to serine) that is uncommon in the proteome. TEV cleavage is efficient at low temperature and can be improved by placing the recognition sequence between two domains (as

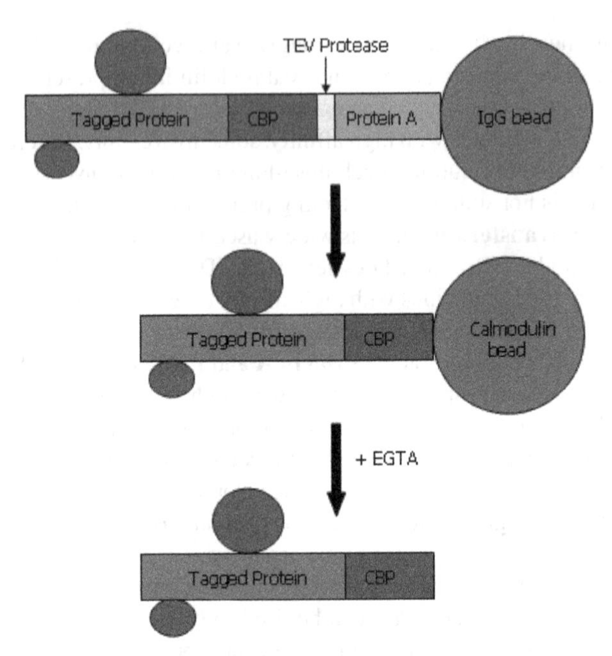

Fig. 1.6 Tandem affinity purification (TAP) coupled to mass spectrometry. Using the TAP approach, a protein of interest (*green*) is expressed as a fusion with two affinity tags separated by a protease cleavage site (here calmodulin-binding protein, tobacco etch virus protease, and Protein A). Associated proteins are represented by *blue* spheres

in the case of the TAP tag method). Other protease cleavage sites may be used. **Enterokinase** recognizes the sequence DDDDKX and cleaves C-terminal to the lysine, although some non-specific cleavage occurs at alternative sites with low frequency. Note that the FLAG tag contains an enterokinase recognition site. **Factor Xa** can also function at low temperatures and cleaves C-terminal to the sequence IEGR. In recent years, inteins have also been successfully used for self-cleaving tagged protein release without the need for protease (Xu et al. 2000).

1.4.3 Genetics and Cloning of Affinity Tagged Proteins

Nearly all affinity tags currently used for high-throughput protein-protein interaction studies are introduced via expression cloning, the DNA encoding the tag being inserted genetically at some point in the gene encoding the bait. Both amino- and carboxy-terminal tags are commonly used. However, each protein is obviously unique, and alternative tagging sites may need to be examined. In some cases, tags may be inserted into non-terminal regions, or into mutant proteins or proteins with regions deleted. In many cases, structural or functional information may help. For example, in cases where post-translational cleavage of the amino terminus generates the mature protein, or where the carboxy-terminus contributes a structural fold essential to function, then these positions would be avoided as sites for introducing an affinity tag.

Large-scale studies are dependent on methods to introduce the DNA into specific genomic locations in a high-throughput manner, and so to date have been most common in model organisms for where such methods are available, notably yeast. Both *Saccharomyces cerevisiae* and *Schizosaccharomyces pombe* contain high efficiency recombination machinery that permits introduction of exogenous DNA in a sequence-specific manner and a number of very powerful genome-wide resources have been generated. Sets of *S. cerevisiae* strains, each encoding an individual gene tagged with GST, GFP, TAP (Protein A-TEV-calmodulin-binding peptide), or FLAG have been generated (Bader et al. 2003). Similar sets comprise strains in which each single gene has been replaced with a marker. The *Escherchia coli* genome has also been extensively tagged (Butland et al. 2005; Arifuzzaman et al. 2006), while a smaller yet significant number of human genes have also been tagged (e.g. (Bouwmeester et al. 2004)). Most of these strains are publicly available.

1.4.4 Isolation of Protein Complexes

Generally, a strain or cell line containing the bait fused to an affinity tag is grown and the cells are lyzed using methods appropriate to the organism. (An exception is where the fusion proteins are generated by in vitro methods such as cell-free translation, where no lysis is necessary). It is important to ensure that the methods are consistent with maintaining the protein-protein interactions. Furthermore, because lysis can lead to mixing of cellular compartments, care should be taken to avoid exposure to proteinase or other enzymes that might degrade proteins or disrupt their interactions. For this reason, proteinase inhibitors and low temperatures are routinely used, and early steps in the procedure (when proteolysis is most likely) should be carried out as rapidly as possible. Affinity-tagged proteins may bind either to media packed into chromatography columns through which lysate is passed, or to media suspended in the lysate. The choice of approach depends on issues such as protein abundance, binding affinity, and other factors like automation or cost. Ideally, the procedure will be optimized for each individual bait. However, in high-throughput projects, this is often impossible, so compromise conditions compatible with the other elements of the project are used.

1.4.5 Proteomics by Mass Spectrometry

Mass spectrometry (MS) is the study of gas phase ions as a means to characterize molecular structure (Aebersold and Mann 2003). Mass spectrometers separate the ions in space or time based on their mass-to-charge ratio (m/z). Currently, proteomics relies especially on two ionization techniques, electrospray ionization (ESI) and matrix-assisted laser desorption ionization (MALDI). ESI involves ionization of peptides at atmospheric pressure by nebulizing a stream of solvent under a potential difference of several thousand volts. The technique, which is usually coupled with triple quadrupole and ion trap or time of flight detectors, can determine protein masses in excess of 150,000 to an accuracy of 0.005%. Nanoelectrospray is a refinement of ESI in which miniaturization of the electrospray source increases the sensitivity of the analysis to the low femtomolar range.

For identification purposes, protein mixtures are typically digested with trypsin before analysis by MS. Peptides resulting from trypsin treatment, which primarily cleaves at lysine and arginine residues that occur approximately every 10–15 residues in proteins, often have size and charge properties that render them effective candidates for ionization. In MALDI, the peptides are embedded in a matrix that absorbs laser light, allowing desorption of ions and analysis by the mass spectrometer. MALDI is most often used in conjunction with time-of-flight (TOF) mass spectrometers, which use transit time differences through a drift region of the instrument to separate ions of different m/z. MALDI-TOF MS permits very sensitive and accurate measurement of peptides up to about 500 kDa. These peptides are generally identified using peptide mass fingerprinting, a technique that compares the peptide masses observed by MS to a set of masses predicted from an in silico digest of proteins encoded by the DNA sequences from genomic databases. Computer programs search for matches between actual and theoretical fragment masses, with strong matches leading to identification of the protein under investigation.

Tandem MS (or MS/MS) is also frequently used in proteomics, particularly with ion trap and quadrupole instruments. In this approach, two or more stages of mass analysis are conducted sequentially. Introduction of an inert gas at the position of the second and/or subsequent mass analyzers causes fragmentation of the initial ionized peptides to produce daughter ions. The product ion spectra can be interpreted to deduce the amino acid sequence of a protein by comparison with predicted patterns obtained from translated protein databases (as with peptide mass fingerprinting). The development of these algorithms was a major advance because it removed the need to manually interpret each mass spectrum and so opened the door to truly high-throughput proteomics.

1.4.6 Identifying Interacting Proteins Using Mass Spectrometry

Both MALDI and MS/MS are commonly used to identify proteins purified using affinity tagging (or other) strategies. Such purifications may vary widely in sample complexity and dynamic range, so protein mixtures may be directly analyzed by MS, or may require some fractionation before or during MS analysis to reduce the complexity of the mixtures entering the mass spectrometers. For example, purified protein preparations may undergo electrophoresis on an SDS polyacrylamide gel (PAGE), stained with silver or Coomassie blue dye, and the visible bands removed and identified by peptide mass fingerprinting using MALDI-TOF MS (Fig. 1.7). Alternatively, an aliquot of purified protein preparation can be digested directly following the purification experiment and the peptides separated by online HPLC (liquid chromatography-tandem mass spectrometry, LC/MS/MS). Advantages of the SDS-PAGE MALDI approach include the fact that the identification is linked to a

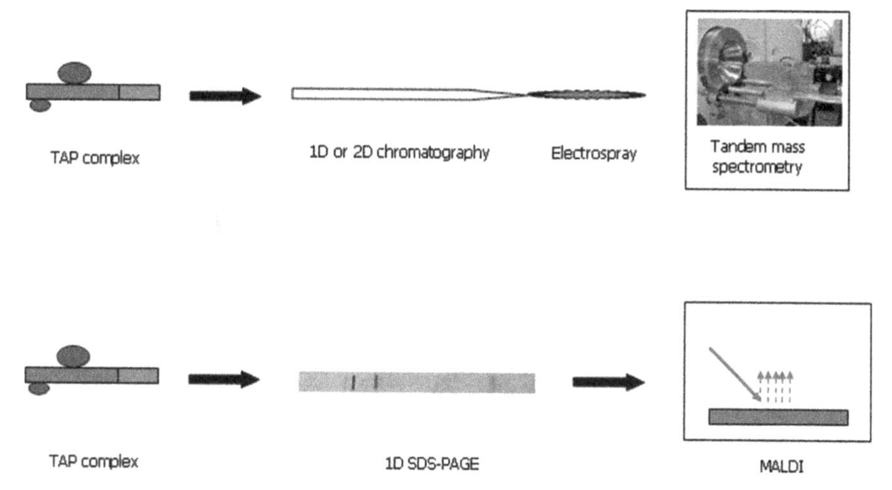

Fig. 1.7 Protein identification by mass spectrometry (MS). The purified TAP complex of proteins can be digested using trypsin and the resulting peptides introduced directly to a mass spectrometer by electrospray ionization following separation by one- or two-dimensional chromatography (*top*). Alternatively, the intact proteins may be separated by gel electrophoresis and the individual bands digested using trypsin before introduction to a mass spectrometer using the MALDI process. Both approaches result in mass spectra that can be used to identify the protein components of the purified mixtures

specific band on a gel (that can be compared to a theoretical protein mass), while advantages of the LC/MS/MS method include less manual effort and in some cases greater sensitivity.

1.4.7 Quantitative Proteomics

Although gels can be used to estimate relative protein abundance differences between samples (using staining intensity), this information is lost upon analysis by LC/MS. Quantitative proteomics data are useful for protein interaction studies because they can help to distinguish "true" interacting proteins from the background of non-specific proteins identified in control experiments. For example, a purified complex should contain all proteins in stoichiometric amounts; quantitivative measurements can determine the stoichiometry and thus tell "true" components of a complex from contaminants. Several approaches have been used to obtain quantitative data, mostly involving the use of stable isotopes such as ^{2}H, ^{13}C or ^{15}N. Proteins or peptides obtained from two or more sources, each labeled so that the equivalent peptides have different masses, can be tracked by mass spectrometry using these mass differences. Quantitative proteomics strategies can be classed as either "pre-experiment" labeling, or "post-experiment" labeling.

In pre-experiment strategies, the isotope is introduced during culture of the organism or cell line. For example, one sample may be grown in normal media while another is grown using media containing deuterated leucine (Ong et al. 2002). Peptides containing deuterated leucine will contain mass shifts relative to those

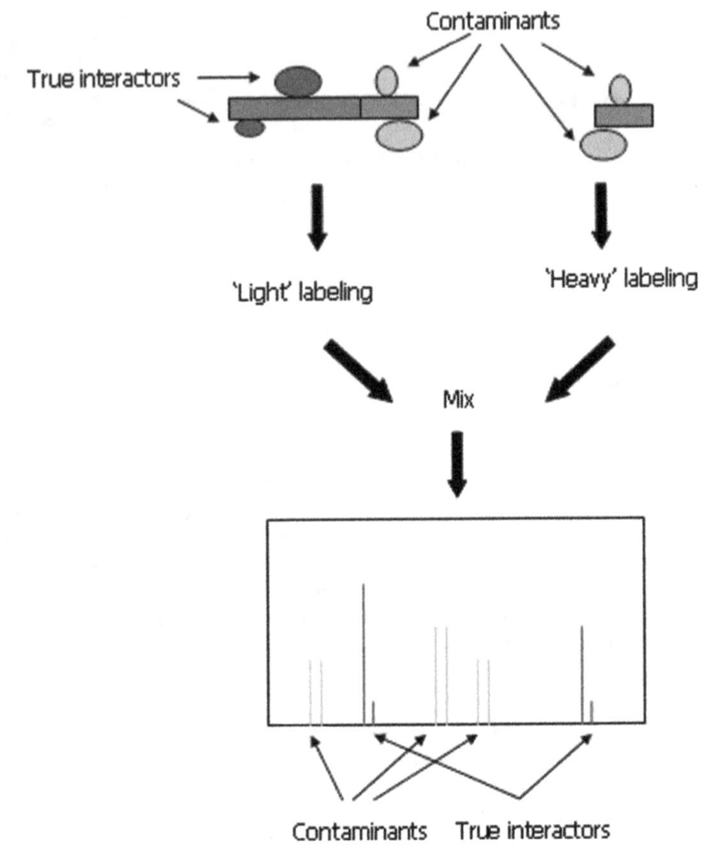

Fig. 1.8 Quantitative proteomics to assess specificity in pulldown experiments. If peptides derived from a control pulldown are labeled using a "heavy" stable isotope-containing reagent, while those from the sample pulldown are labeled using a "light" reagent, true interactors can be distinguished from contaminant proteins. This is because the heavy and light contaminant peaks are derived from both samples in approximately equal amounts, while light peaks will be significantly more intense than light peaks for peptides that are only present in the true sample fraction

containing natural leucine and the relative signal emanating from each ion in the mass spectrometer can be used to estimate the relative abundance of the parent peptides. In post-experiment labeling, the label is introduced chemically or enzymatically after the proteins have been purified (Fig. 1.8). Several variants have been described. The isotope-coded affinity tag (ICAT) method described by Gygi and coworkers (Gygi et al. 1999) uses a biotinylated thiol-active reagent containing a linker that is produced in alternative heavy (containing eight deuterium atoms) and light (containing eight hydrogen atoms) forms. The thiol-active component of the reagent binds cysteine-containing peptides in the lysate. When comparing two protein samples, both forms are used, so cysteinylated peptides in one sample are heavy-labelled while the corresponding peptides in the other sample are light-labelled. The two fractions are pooled, purified by solid phase extraction, and analyzed by LCMS. The heavy-light pairs for each peptide can be recognized by their mass difference during total ion scans, and the relative peak area for these peaks is used to determine the relative abundance of the corresponding peptides in the original sample.

1.5 Far-Western Blotting

Western blotting can also be used to detect protein-protein interactions. Instead of detecting a blotted protein on a membrane using antibodies, the membrane can be incubated first with another protein that binds to the membrane-bound protein (Fig. 1.9). After a wash step the membrane is incubated with an antibody against the secondary protein which is only detected if it bound to the primary protein on the membrane. If there is no antibody available against the secondary protein, epitope-tagged proteins are used.

Far-Western Blotting is pretty much identical to a regular Western blot except an additional incubation which adds another layer of protein. This is why this method is called "far Western" blotting. The method is usually used to confirm protein interactions detected by yeast two-hybrid screens and other screening technologies.

Far Western blotting can also be used to map interaction domains. A protein of interest is first partially digested with a protease and the fragments separated on a gel and then blotted onto a membrane. The membrane is then incubated with an interacting protein which binds only to the fragments that contain the interaction domain. The primary protein can be labelled N- or C-terminally to detect all fragments that contain the N-or C-terminus. From the size of the bands on the Western blot the fragment with the interaction domain can often be inferred.

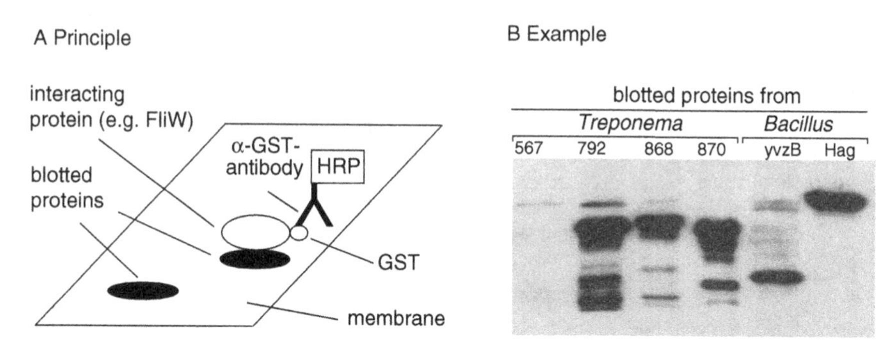

Fig. 1.9 Far-Western blot. (**A**) General principle (see text for details). (**B**) Example showing that protein FliW (*Bacillus subtilis*) binds to flagellar proteins of several bacteria. The proteins TP0567 (negative control), TP00792 (=FlaB2), TP0868 (= FlaB1), TP0870 (= FlaB3) of *Treponema pallidum*, the causative agent of syphilis, were expressed together with flagellar proteins YvzB and Hag from *Bacillus subtilis* in *E. coli* and an extract from the latter blotted onto a membrane. This membrane was then incubated with a GST-FliW fusion protein (where FliW was from Bacillus subtilis this time) and the fusion protein detected with an anti-GST antibody that was coupled to horse-radish peroxidase (HRP) which emits light when its substrate is added. This far Western not only shows that FliW binds to flagellar proteins FlaB1-3 but also that this interaction is conserved in multiple, only distantly related bacteria, namely *Treponema* und *Bacillus*. Modified after (Titz et al. 2006)

1.6 Protein and Peptide Chips

Proteins can also be printed onto glass such as microscope slides and these protein chips can be then screened with labelled proteins. Zhu et al. (2001) were the first to make a proteome-wide chip with almost all GST-His6-tagged yeast proteins attached on a single slide that was coated with Nickel (Zhu et al. 2001). They screened these slides with biotinylated calmodulin and detected 6 previously known Calmodulin interactors and 33 new ones. So far, not many additional screens have been published and thus it is difficult to judge how protein chips do in comparison with other methods.

In addition to full-length proteins, glass slides or membranes can also be coated with short peptides, usually of 10–30 amino acids in length. Such peptide arrays can be screened for interacting proteins when labelled proteins are incubated with them, similar to protein chips. Their main application is the mapping of interaction epitopes (Fig. 1.10). This is based on the fact that most proteins bind to relatively short linear peptides. Since the membrane-bound peptides are not sterically constrained as when they are in a folded protein, they can be induced to fit into a protein that is used to screen the library. However, this flexibility also may lead to false positives.

1.7 Quality of Large-Scale Interaction Data

Although several methods for the detection of protein-protein interactions exist, no method is able to identify all protein-protein interactions – each experimental strategy generates a significant number of **false negatives**. The reasons for this systematic error are only partly understood. Two-hybrid false negatives might be

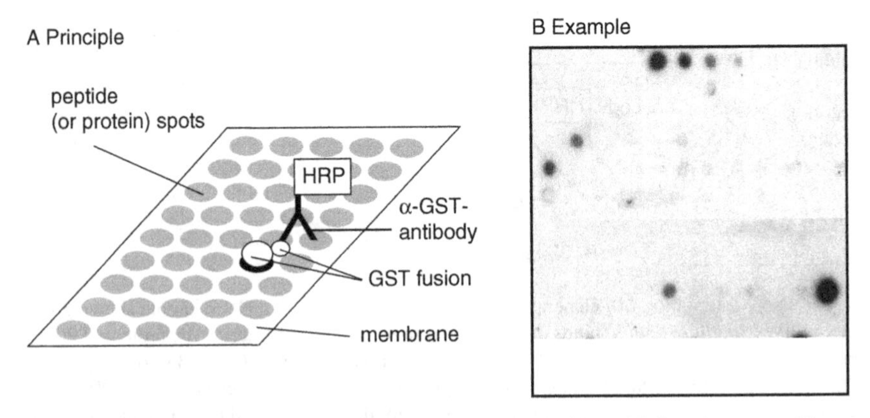

Fig. 1.10 Peptide and protein arrays are similar to far-Western blots. (**A**) Proteins or peptides are spotted on membranes and detected by interacting proteins and pertinent antibodies. Peptides can also be synthesized directly on the membrane using a technology called SPOT synthesis. (**B**) Here, peptides were synthesized by SPOT-synthesis and probed with GST-fusions of an FF-domain from yeast. Courtesy of Claudia Ester

caused by insufficient expression or nuclear localization of the tested proteins, by sterical effects due to the usage of two fusion proteins ("two-hybrid"), or involve weak interactions within complexes that require cooperative effects to be stabilized (Aloy and Russell 2002). Mass spectrometry analysis, on the other hand, often has problems with low abundance proteins and proteins that are only weakly associated with protein complexes (or transient interactions) and hence tend to get lost during purification.

False negatives lead to gaps in our picture of the internal structure of the cell. However, more serious are **false positive** interactions, which result in erroneous data and thus misleading conclusions. Two types of false positives need to be distinguished: technical false positives and biological false positives. In yeast-two-hybrid studies, technical false positives can arise by bait constructs, which activate the reporter gene without interacting with a prey ("self activating baits"). In addition, mutations in the reporter genes or incorrect folding in the unnatural environment are sources for technical false positives. On contrary, "biological false positives" represent true interactions that take place in the Y2H system but have no biological relevance (Ito et al. 2002). Examples are proteins that are interacting in the Y2H system but are expressed in different cell types or different organelles in vivo.

A number of studies tried to estimate the number of true positives for high-throughput interaction studies. The critical point of any attempt to estimate the number of true and false positives in a HTS interaction study is the choice of the "true positive" data set against which the new interactions are evaluated.

To estimate the overall interaction reliability, Deane et al. compared the co-expression profiles of known interacting proteins with protein pairs from high-throughput screens (Deane et al. 2002). Based on this comparison, they estimate a false-positive rate of 50–70% for yeast-two-hybrid experiments. Sprinzak et al. tried to estimate the interaction quality based on the observed degree of co-localization and shared functional role of the interacting proteins (Sprinzak et al. 2003). This estimation yielded a false-positive rate of ~50% for large-scale yeast-two-hybrid studies. Patil et al. used a combination of three genomic features (known interacting domains, gene annotations, and sequence homology) with a Bayesian network approach (Patil and Nakamura 2005) and estimate that 56% of the high-throughput interactions for yeast have high reliability.

Edwards et al. (2002) selected known interactions from 3D-structures (RNA polymerase II, proteasome and the Arp2/3 complex), and additionally, complexes from the literature. The crystal structures of complexes approximate the "absolute truth" about stable protein interactions because they reveal all interactions in atomic detail, at least for the proteins that have been co-crystallized. Based on crystal structures, Edwards et al. found a false negative rate of 51–96% for yeast-two-hybrid and of 15–50% for in vivo pull-down experiments, respectively. In this context it is remarkable that conventional "low throughput" methods also produce a large fraction of false positives – for example 61% in a pull-down study of RNA polymerase II (Edwards et al. 2002). Another method proposed by D'Haeseleer and Church (2004) does not rely on a gold-standard and involves comparing two interaction datasets to each other and to a reference dataset. In this study, a false positive rate between ~50–90% (depending on the dataset) was calculated both for Y2H and for coAP/MS experiments.

Overall, several approaches to estimate the reliability of two-hybrid interactions conclude that 50% or more are true positive interactions. This is underlined by the finding in recent large-scale yeast-two hybrid studies that between 50% and 70% of the identified interactions can be reproduced by an independent method (which also has a certain false-negative rate) (Rual et al. 2005; Uetz et al. 2006).

However, several approaches were devised to minimize the number of false positives further. These approaches rely either on the identification of intrinsic properties, which lead to unspecific interactions, or on the integration of several datasets (and data types). Uetz et al. (2000) systematically evaluated the signal/background ratio and discarded yeast-two-hybrid interactions, which could not be reproduced. Ito et al. (2001) defined interacting protein pairs found three or more times as the (supposedly reliable) "core" dataset. Rain et al. screened bait proteins against a genomic fragment prey library and considered overlapping prey fragments as the most reliable. This approach combines reproducibility and identifies the interacting domain at the same time (Rain et al. 2001). Giot et al. employed a statistical technique (logistic regression) together with a set of known "gold standard" interaction to identify properties of true positive interactions. The authors estimated their filtered high confidence network to retain 40% interactions of biological significance (Giot et al. 2003).

All these approaches are based on the evaluation of intrinsic properties of the respective system. Further evidence for reliable interactions can be obtained by integration of several data sources and data types. Several studies showed that interacting proteins tend to be co-expressed at the mRNA level under various experimental conditions (Jansen et al. 2002) (Ge et al. 2001; Grigoriev 2001). However, while co-expression of the two partners increases the confidence in a protein-protein interaction, it is only an indirect measure of its reliability. Proteins in a complex may need to be expressed at similar levels in order to maintain their stochiometric ratios, but this is certainly not true for all complexes and even less so for transient interactions that are often found in Y2H screens. Reliability can be also gained by looking at the interactions of paralogous proteins. Interactions reproduced with paralogous proteins were labeled has highly reliable by Deane et al. (2002). However, many proteins do not have paralogs and paralogs with diversified functions do not need to retain the same (or even similar) interactions.

1.8 Comparison of Methods

1.8.1 Y2H vs. co-AP/MS

Detection of protein interactions by Y2H and coAP/MS differ in a number of important aspects (Fig. 1.11). First, Y2H assays mainly detect direct binary interactions, whereas coAP/MS detects one-to-many relationships, which usually also include

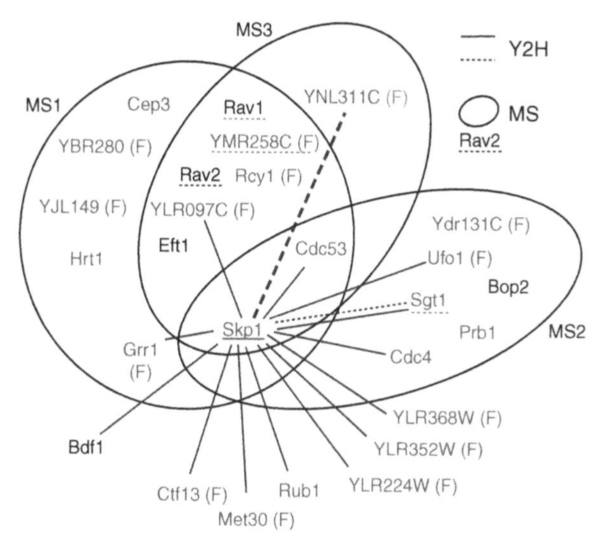

Fig. 1.11 Comparison of interaction data gained by Y2H and MS. Skp1 is a protein involved in ubiquitin-mediated protein degradation and has been epitope-tagged for both Y2H screens and MS analysis. The purified complexes of Skp1 from three independent MS studies (*circles*, MS1-3) and the binary interactions from two Y2H studies (*straigth* and *dashed lines*) are shown. Despite the differences in the data sets, most of the discovered interactions seem to be plausible: most proteins are known to be involved in protein degradation. Skp1 is directed to its target proteins via so-called F-box proteins, which contain a short peptide motif, the F-box (F). Note that neither Gavin et al. (2006) nor Krogan et al. (2006) found any proteins with Skp1 as bait. However, Gavin et al. found Skp1 in purifications with Rav2, Ymr258c, and Ypt52 as bait and Krogan et al. found Skp1 with 12 bait proteins, namely Cdc34, Csn9, Gcd11, Pcl8, Pol30, Rav1, Ril1, Sgt1, Vip1, Vma2, and Ymr258c. Proteins found in these purifcations are indicated by dashed underlining. Modified after (Titz et al. 2004)

indirectly interacting proteins. Second, Y2H analysis is often conducted under non-physiological expression conditions, whereas for coAP/MS, detection can be done at endogenous expression levels. Third, because of the previous reason, Y2H interactions are hypothetical interactions which may not take place in vivo or only under conditions that are not yet known. Purified complexes are forming under the condition under which the complex was purified. However, different complexes may form under different conditions and laboratory conditions may not even reflect the normal circumstances in the life of a cell.

The published experimental approaches for high-throughput interaction analyses employing these methods have taught us already one important lesson: Y2H and MS datasets are strikingly different but also complementary. This difference between datasets – even between datasets derived by a similar method for the same species – is exemplified by two recent coAP/MS studies for yeast (Gavin et al. 2006; Krogan et al. 2006). Goll and Uetz found, that only 28% of the core complexes from Gavin et al. are completely contained in the complexes identified by Krogan et al. Only 6 out of about 500 complexes were identical (Goll and Uetz 2006).

The complementary nature of Y2H and coAP/MS methods seems to depend on molecular properties of the interactions: Transient interactions are often found by yeast-two-hybrid analysis but not AP, whereas stable interactions (such as those in protein complexes) are more reliably identified by in vivo pull-down techniques (Aloy and Russell 2002). This finding is not surprising, given the highly cooperative forces that stabilize a protein complex: weak interactions in a complex will not be detected by Y2H analysis as long as only pairs of proteins are tested that are not stabilized by the other subunits of a complex. In addition, coAP/MS methods are biased towards highly abundant proteins (von Mering et al. 2002). Y2H assays are less biased in this respect as they use proteins expressed at sufficient levels from a heterologous promoter.

1.8.2 coAP/MS vs Protein Chips

In the first large-scale ErbB-interactome study, Schulze et al. (2005) used a quantitative proteomic approach for identifying proteins associated with phosphotyrosine motifs of the ErbB receptor members (Schulze et al. 2005). To this end, they generated 89 different "bait" peptides in phosphorylated and unphosphorylated forms covering all intracellular tyrosine residues of the ErbB-family members. In the next step, these peptides bearing phosphorylated and unphosphorylated tyrosine residues were incubated with HeLa cell lysates and pulldown experiments were performed to enrich specific binding partners for the phosphorylated bait peptides. Lastly, these proteins were identified and quantified by mass spectrometry. Surprisingly, only 40 out of the 89 examined tyrosine residues did have an interacting protein in their phosphorylated form. Most of the tyrosine residues that interacted with specific partners accumulated at the C-terminal regions of the receptors. This study also indicated that the distribution of interacting partners of the different ErbB members shows clear differences between individual receptors, but also a significant overlap. For example, both EGFR and ErbB4 have multiple binding sites for the adaptor proteins Shc and Grb2, but only EGFR binds the ubiquitin ligase Cbl, whereas ErbB4 is unique in binding Nck. Altogether, Schulze et al found 10 ErbB interactors, all of which have either an SH2 or PTB domain, thus emphasizing the specificity of these signal transduction modules (Pawson and Nash 2003).

In a related project, Jones et al used protein microarrays to identify ErbB interactors (Jones et al. 2006). Since the phosphotyrosines in ErbB receptors are primarily bound by SH2 and PTB domains, Jones et al successfully expressed and purified 106 (out of 109) SH2 domains and 41 (out of 44) PTB domains encoded in the human genome. These domains were then spotted onto glass slides and subsequently probed with ErbB peptides that contained phosphorylated and nonphosphorylated tyrosines, respectively. While Schulze et al used all possible 89 tyrosine-containing ErbB peptides for their pulldowns, Jones et al concentrated on 29 peptides that were known to be phosphorylated in EGFR, ErbB2 and ErbB3, and four peptides that were predicted to be phosphorylated in ErbB4. These experiments revealed that each phosphotyrosine on EGFR binds seven different proteins on average, whereas those on ErbB2, ErbB3 and ErbB4 bind 17, 9 and 2 proteins on average, respectively. This adds up to many interactions, namely 54 with EGFR, 59 with ErbB2, 37 with ErbB3 and 8 with ErbB4, most of which were new. The sobering fact is that the mass spec study by Schulze et al found quite different, and, in fact, much smaller numbers, namely nine, four, four, and eight different interacting proteins for the four receptors, respectively. While at least the number of identified ErbB4 interactors by both approaches appears to be identical, only one protein (Shc) is actually common to both ErbB4 data sets (Uetz and Stagljar 2006).

Where do these dramatic differences come from? Well, simply from the fact that the two groups looked at very different things: while Schulze et al. pulled down proteins that most likely bind to ErbB receptors in HeLa cells, Jones et al looked at the whole SH2/PTB interaction space of ErbB receptors in vitro. That is, Jones et al told us which SH2 and PTB domains may bind to which receptor if both are present in a cell. In contrast, Schulze et al told us which SH2 and PTB proteins bind to ErbB receptors specifically in HeLa cells. Unfortunately, we have to wait until further studies reveal which proteins can be pulled down in the other 200+ human cell types or, alternatively, which of the ErbB receptors and SH2/PTB proteins are expressed together in those cells. At least we know that the proteins detected by Schulze are coexpressed in HeLa; some differences (e.g., Grb2 interacts with all four ErbB in Schulze but only with ErbB2 in Jones; STAT5 interaction is missed altogether in Jones) might also be due to possible methodological differences.

1.9 Conclusions

It should have become clear from this chapter that none of the technologies available today is perfect. No technology can identify all interactions, and all of them have a certain fraction of false positives and false negatives. Furthermore, quantitative technologies are only being developed and none of them has been applied on a large scale. Computational methods are required to assess the quality of interaction data and thus to prioritize them for further study. One way to do this is to integrate various datasets including non-interaction data such as structural information. The other chapters in this book will detail some of these strategies and tools.

Glossary of Abbreviations

5-FOA fluoro-orotic acid, a chemical that inhibits the HIS3 enzyme often used in the yeast two-hybrid system.

AD (transcriptional) activation domain.

B2H Bacterial two-hybrid

B42 artifical bacterial transcriptional activator; used as an alternative to the yeast GAL4-AD.

CBD Cellulose-binding domain or Chitin-binding domain

CBP calmodulin-binding peptide (CBP) binds calmodulin in the presence of calcium, and the interaction can be disrupted by EGTA.

Co-AP co-affinity purification

Co-IP co-immunoprecipitation (see text for details)

DBD DNA-binding domain

EGFR epidermal growth factor receptor

EGTA ethylene glycol tetraacetic acid, a chelating agent that is related to the better known EDTA, but with a much higher affinity for calcium than for magnesium ions.

ESI electrospray ionization (ESI)

FC fragment complementation

FRET fluorescent resonance energy transfer

GAL4 a yeast transcription factor that served as basis for the first Y2H assay which used the DBD and AD of this protein. GAL4 regulates genes that are involved in galactose metabolism.

GFP Green fluorescent protein

GST Glutathion-S-Transferase

HA hemagglutinin, a protein and antigen of Influenza virus.

HIS3 Imidazoleglycerolphosphate (IGP) dehydratase, an enzyme that catalyzes the seventh step in the histidine biosynthesis pathway

HRP horse-radish peroxidase

HTS high-throughput screening

ICAT isotope-coded affinity tag, a chemical tag used to label proteins for subsequent MS analysis.

LacZ Beta-galactosidase. This enzyme from *E. coli* is often used as reporter gene in Y2H assays.

LCMS liquid chromatography mass spectrometry

LEU2 3-Isopropylmalate dehydrogenase, an enzyme that catalyzes the third step in the leucine biosynthesis pathway.

LexA bacterial repressor and DNA-binding protein that is used as an alternative to GAL4.

M2H mammalian two-hybrid

MALDI matrix-assisted laser desorption ionization

MBD Maltose-binding protein

MS Mass spectrometry.

PBS phosphate-buffered saline, a pH-stabilized buffer solution.

PTB phospho-tyrosine-binding domain

SAI socio-affinity index, a measure that describes the tendency of a protein to interact specifically with others

SDS-PAGE sodium dodecyl sulphate poly-acrylamide gel electrophoresis

SH2 Src homology domain 2

SH3 Src homology domain 3

TAP Tandem Affinity Purification

TEV tobacco etch virus

TOF time of flight, usually used in combination with MALDI ("MALDI-TOF"), a special kind of mass spectrometer.

URA3 Orotidine-5'-phosphate decarboxylase, an enzyme that catalyzes the sixth step of pyrimidine biosynthesis

Y2H Yeast two-hybrid system

References

Aebersold, R. and Mann, M. (2003). Mass spectrometry-based proteomics. Nature 422(6928): 198–207.

Aloy, P. and Russell, R. B. (2002). The third dimension for protein interactions and complexes. Trends Biochem. Sci. 27(12): 633–638.

Arifuzzaman, M., Maeda, M., Itoh, A., Nishikata, K., Takita, C., Saito, R., Ara, T., Nakahigashi, K., Huang, H. C., Hirai, A., Tsuzuki, K., Nakamura, S., Altaf-Ul-Amin, M., Oshima, T., Baba, T., Yamamoto, N., Kawamura, T., Ioka-Nakamichi, T., Kitagawa, M., Tomita, M., Kanaya, S., Wada, C. and Mori, H. (2006). Large-scale identification of protein-protein interaction of Escherichia coli K-12. Genome Res. 16(5): 686–691.

Aronheim, A. (1997). Improved efficiency sos recruitment system: expression of the mammalian GAP reduces isolation of Ras GTPase false positives. Nucleic Acids Res. 25(16): 3373–3374.

Bader, G. D., Heilbut, A., Andrews, B., Tyers, M., Hughes, T. and Boone, C. (2003). Functional genomics and proteomics: charting a multidimensional map of the yeast cell. Trends Cell Biol. 13(7): 344–356.

Bouwmeester, T., Bauch, A., Ruffner, H., Angrand, P. O., Bergamini, G., Croughton, K., Cruciat, C., Eberhard, D., Gagneur, J., Ghidelli, S., Hopf, C., Huhse, B., Mangano, R., Michon, A. M., Schirle, M., Schlegl, J., Schwab, M., Stein, M. A., Bauer, A., Casari, G., Drewes, G., Gavin, A. C., Jackson, D. B., Joberty, G., Neubauer, G., Rick, J., Kuster, B. and Superti-Furga, G. (2004). A physical and functional map of the human TNF-alpha/NF-kappa B signal transduction pathway. Nat. Cell Biol. 6(2): 97–105.

Butland, G., Peregrin-Alvarez, J. M., Li, J., Yang, W., Yang, X., Canadien, V., Starostine, A., Richards, D., Beattie, B., Krogan, N., Davey, M., Parkinson, J., Greenblatt, J. and Emili, A. (2005). Interaction network containing conserved and essential protein complexes in Escherichia coli. Nature 433(7025): 531–537.

Cagney, G., Uetz, P. and Fields, S. (2000). High-throughput screening for protein-protein interactions using two-hybrid assay. Methods Enzymol. 328: 3–14.

D'Haeseleer, P. and Church, G. M. (2004). Estimating and improving protein interaction error rates. Proc. IEEE Comput. Syst. Bioinform. Conf.: 216–223.

Deane, C. M., Salwinski, L., Xenarios, I. and Eisenberg, D. (2002). Protein interactions: two methods for assessment of the reliability of high throughput observations. Mol. Cell. Proteomics 1(5): 349–356.

Edwards, A. M., Kus, B., Jansen, R., Greenbaum, D., Greenblatt, J. and Gerstein, M. (2002). Bridging structural biology and genomics: assessing protein interaction data with known complexes. Trends Genet. 18(10): 529–536.

Eyckerman, S., Verhee, A., der Heyden, J. V., Lemmens, I., Ostade, X. V., Vandekerckhove, J. and Tavernier, J. (2001). Design and application of a cytokine-receptor-based interaction trap. Nat. Cell Biol. 3(12): 1114–1119.

Fields, S. and Song, O. (1989). A novel genetic system to detect protein-protein interactions. Nature 340(6230): 245–246.

Gavin, A. C., Aloy, P., Grandi, P., Krause, R., Boesche, M., Marzioch, M., Rau, C., Jensen, L. J., Bastuck, S., Dumpelfeld, B., Edelmann, A., Heurtier, M. A., Hoffman, V., Hoefert, C., Klein, K., Hudak, M., Michon, A. M., Schelder, M., Schirle, M., Remor, M., Rudi, T., Hooper, S., Bauer, A., Bouwmeester, T., Casari, G., Drewes, G., Neubauer, G., Rick, J. M., Kuster, B., Bork, P., Russell, R. B. and Superti-Furga, G. (2006). Proteome survey reveals modularity of the yeast cell machinery. Nature 440(7084): 631–636.

Ge, H., Liu, Z., Church, G. M. and Vidal, M. (2001). Correlation between transcriptome and interactome mapping data from Saccharomyces cerevisiae. Nat. Genet. 29(4): 482–486.

Giot, L., Bader, J. S., Brouwer, C., Chaudhuri, A., Kuang, B., Li, Y., Hao, Y. L., Ooi, C. E., Godwin, B., Vitols, E., Vijayadamodar, G., Pochart, P., Machineni, H., Welsh, M., Kong, Y., Zerhusen, B., Malcolm, R., Varrone, Z., Collis, A., Minto, M., Burgess, S., McDaniel, L., Stimpson, E., Spriggs, F., Williams, J., Neurath, K., Ioime, N., Agee, M., Voss, E., Furtak, K., Renzulli, R., Aanensen, N., Carrolla, S., Bickelhaupt, E., Lazovatsky, Y., DaSilva, A., Zhong, J., Stanyon, C. A., Finley, R. L., Jr., White, K. P., Braverman, M., Jarvie, T., Gold, S., Leach, M., Knight, J., Shimkets, R. A., McKenna, M. P., Chant, J. and Rothberg, J. M. (2003). A protein interaction map of Drosophila melanogaster. Science 302(5651): 1727–1736.

Golemis, E. A. and Khazak, V. (1997). Alternative yeast two-hybrid systems. The interaction trap and interaction mating. Methods Mol. Biol. 63: 197–218.

Goll, J. and Uetz, P. (2006). The elusive yeast interactome. Genome Biol. 7(6): 214.211–216.

Grigoriev, A. (2001). A relationship between gene expression and protein interactions on the proteome scale: analysis of the bacteriophage T7 and the yeast Saccharomyces cerevisiae. Nucleic Acids Res. 29(17): 3513–3519.

Gygi, S. P., Rist, B., Gerber, S. A., Turecek, F., Gelb, M. H. and Aebersold, R. (1999). Quantitative analysis of complex protein mixtures using isotope-coded affinity tags. Nat. Biotechnol. 17(10): 994–999.

Hart, G. T., Ramani, A. K. and Marcotte, E. M. (2006). How complete are current yeast and human protein-interaction networks? Genome Biol. 7(11): 120.

Hu, C. D. and Kerppola, T. K. (2003). Simultaneous visualization of multiple protein interactions in living cells using multicolor fluorescence complementation analysis. Nat. Biotechnol. 21(5): 539–545.

Ito, T., Chiba, T., Ozawa, R., Yoshida, M., Hattori, M. and Sakaki, Y. (2001). A comprehensive two-hybrid analysis to explore the yeast protein interactome. Proc. Natl. Acad. Sci. U. S. A. 98(8): 4569–4574.

Ito, T., Ota, K., Kubota, H., Yamaguchi, Y., Chiba, T., Sakuraba, K. and Yoshida, M. (2002). Roles for the two-hybrid system in exploration of the yeast protein interactome. Mol. Cell. Proteomics 1(8): 561–566.

Jansen, R., Lan, N., Qian, J. and Gerstein, M. (2002). Integration of genomic datasets to predict protein complexes in yeast. J. Struct. Funct. Genomics 2(2): 71–81.
Jin, F., Avramova, L., Huang, J. and Hazbun, T. (2007). A yeast two-hybrid smart-pool-array system for protein-interaction mapping. Nat. Methods 4(5): 405–407.
Johnsson, N. and Varshavsky, A. (1994). Split ubiquitin as a sensor of protein interactions in vivo. Proc. Natl. Acad. Sci. U. S. A. 91(22): 10340–10344.
Jones, R. B., Gordus, A., Krall, J. A. and MacBeath, G. (2006). A quantitative protein interaction network for the ErbB receptors using protein microarrays. Nature 439(7073): 168–174.
Joung, J. K., Ramm, E. I. and Pabo, C. O. (2000). A bacterial two-hybrid selection system for studying protein-DNA and protein-protein interactions. Proc. Natl. Acad. Sci. U. S. A. 97(13): 7382–7387.
Karimova, G., Pidoux, J., Ullmann, A. and Ladant, D. (1998). A bacterial two-hybrid system based on a reconstituted signal transduction pathway. Proc. Natl. Acad. Sci. U. S. A. 95(10): 5752–5756.
Krogan, N. J., Cagney, G., Yu, H., Zhong, G., Guo, X., Ignatchenko, A., Li, J., Pu, S., Datta, N., Tikuisis, A. P., Punna, T., Peregrin-Alvarez, J. M., Shales, M., Zhang, X., Davey, M., Robinson, M. D., Paccanaro, A., Bray, J. E., Sheung, A., Beattie, B., Richards, D. P., Canadien, V., Lalev, A., Mena, F., Wong, P., Starostine, A., Canete, M. M., Vlasblom, J., Wu, S., Orsi, C., Collins, S. R., Chandran, S., Haw, R., Rilstone, J. J., Gandi, K., Thompson, N. J., Musso, G., St Onge, P., Ghanny, S., Lam, M. H., Butland, G., Altaf-Ul, A. M., Kanaya, S., Shilatifard, A., O'Shea, E., Weissman, J. S., Ingles, C. J., Hughes, T. R., Parkinson, J., Gerstein, M., Wodak, S. J., Emili, A. and Greenblatt, J. F. (2006). Global landscape of protein complexes in the yeast Saccharomyces cerevisiae. Nature 440(7084): 637–643.
Luo, Y., Batalao, A., Zhou, H. and Zhu, L. (1997). Mammalian two-hybrid system: a complementary approach to the yeast two-hybrid system. Biotechniques 22(2): 350–352.
Marti, F., Xu, C. W., Selvakumar, A., Brent, R., Dupont, B. and King, P. D. (1998). LCK-phosphorylated human killer cell-inhibitory receptors recruit and activate phosphatidylinositol 3-kinase. Proc. Natl. Acad. Sci. U. S. A. 95(20): 11810–11815.
Ong, S. E., Blagoev, B., Kratchmarova, I., Kristensen, D. B., Steen, H., Pandey, A. and Mann, M. (2002). Stable isotope labeling by amino acids in cell culture, SILAC, as a simple and accurate approach to expression proteomics. Mol. Cell. Proteomics 1(5): 376–386.
Patil, A. and Nakamura, H. (2005). Filtering high-throughput protein-protein interaction data using a combination of genomic features. BMC Bioinformatics 6: 100.
Pawson, T. and Nash, P. (2003). Assembly of cell regulatory systems through protein interaction domains. Science 300(5618): 445–452.
Porath, J., Carlsson, J., Olsson, I. and Belfrage, G. (1975). Metal chelate affinity chromatography, a new approach to protein fractionation. Nature 258(5536): 598–599.
Rain, J. C., Selig, L., De Reuse, H., Battaglia, V., Reverdy, C., Simon, S., Lenzen, G., Petel, F., Wojcik, J., Schachter, V., Chemama, Y., Labigne, A. and Legrain, P. (2001). The protein-protein interaction map of Helicobacter pylori. Nature 409(6817): 211–215.
Rigaut, G., Shevchenko, A., Rutz, B., Wilm, M., Mann, M. and Seraphin, B. (1999). A generic protein purification method for protein complex characterization and proteome exploration. Nat. Biotechnol. 17(10): 1030–1032.
Ron, D. and Dressler, H. (1992). pGSTag–a versatile bacterial expression plasmid for enzymatic labeling of recombinant proteins. Biotechniques 13(6): 866–869.
Rual, J. F., Venkatesan, K., Hao, T., Hirozane-Kishikawa, T., Dricot, A., Li, N., Berriz, G. F., Gibbons, F. D., Dreze, M., Ayivi-Guedehoussou, N., Klitgord, N., Simon, C., Boxem, M., Milstein, S., Rosenberg, J., Goldberg, D. S., Zhang, L. V., Wong, S. L., Franklin, G., Li, S., Albala, J. S., Lim, J., Fraughton, C., Llamosas, E., Cevik, S., Bex, C., Lamesch, P., Sikorski, R. S., Vandenhaute, J., Zoghbi, H. Y., Smolyar, A., Bosak, S., Sequerra, R., Doucette-Stamm, L., Cusick, M. E., Hill, D. E., Roth, F. P. and Vidal, M. (2005). Towards a proteome-scale map of the human protein-protein interaction network. Nature 437(7062): 1173–1178.

Schulze, W. X., Deng, L. and Mann, M. (2005). Phosphotyrosine interactome of the ErbB-receptor kinase family. Mol. Syst. Biol. 1: 2005 0008.

Sprinzak, E., Sattath, S. and Margalit, H. (2003). How reliable are experimental protein-protein interaction data? J. Mol. Biol. 327(5): 919–923.

Stagljar, I., Korostensky, C., Johnsson, N. and te Heesen, S. (1998). A genetic system based on split-ubiquitin for the analysis of interactions between membrane proteins in vivo. Proc. Natl. Acad. Sci. U. S. A. 95(9): 5187–5192.

Stelzl, U., Worm, U., Lalowski, M., Haenig, C., Brembeck, F. H., Goehler, H., Stroedicke, M., Zenkner, M., Schoenherr, A., Koeppen, S., Timm, J., Mintzlaff, S., Abraham, C., Bock, N., Kietzmann, S., Goedde, A., Toksoz, E., Droege, A., Krobitsch, S., Korn, B., Birchmeier, W., Lehrach, H. and Wanker, E. E. (2005). A human protein-protein interaction network: a resource for annotating the proteome. Cell 122(6): 957–968.

Terpe, K. (2003). Overview of tag protein fusions: from molecular and biochemical fundamentals to commercial systems. Appl. Microbiol. Biotechnol. 60(5): 523–533.

Titz, B., Rajagopala, S. V., Ester, C., Hauser, R. and Uetz, P. (2006). A novel conserved assembly factor of the bacterial flagellum. J. Bacteriol. 188: 7700–7706.

Titz, B., Schlesner, M. and Uetz, P. (2004). What do we learn from high-throughput protein interaction data? Expert Rev. Proteomics 1(1): 111–121.

Uetz, P., Dong, Y. A., Zeretzke, C., Atzler, C., Baiker, A., Berger, B., Rajagopala, S. V., Roupelieva, M., Rose, D., Fossum, E. and Haas, J. (2006). Herpesviral protein networks and their interaction with the human proteome. Science 311(5758): 239–242.

Uetz, P., Fumagalli, S., James, D. and Zeller, R. (1996). Molecular interaction between limb deformity proteins (formins) and Src family kinases. J. Biol. Chem. 271(52): 33525–33530.

Uetz, P., Giot, L., Cagney, G., Mansfield, T. A., Judson, R. S., Knight, J. R., Lockshon, D., Narayan, V., Srinivasan, M., Pochart, P., Qureshi-Emili, A., Li, Y., Godwin, B., Conover, D., Kalbfleisch, T., Vijayadamodar, G., Yang, M., Johnston, M., Fields, S. and Rothberg, J. M. (2000). A comprehensive analysis of protein-protein interactions in Saccharomyces cerevisiae. Nature 403(6770): 623–627.

Uetz, P. and Stagljar, I. (2006). The interactome of human EGF/ErbB receptors. Mol. Syst. Biol. 2: 2006 0006.

Vojtek, A. B., Hollenberg, S. M. and Cooper, J. A. (1993). Mammalian Ras interacts directly with the serine/threonine kinase Raf. Cell 74(1): 205–214.

Vollert, C. S. and Uetz, P. (2004). The phox homology (PX) domain protein interaction network in yeast. Mol. Cell. Proteomics 3(11): 1053–1064.

von Mering, C., Krause, R., Snel, B., Cornell, M., Oliver, S. G., Fields, S. and Bork, P. (2002). Comparative assessment of large-scale data sets of protein-protein interactions. Nature 417(6887): 399–403.

Wehr, M. C., Laage, R., Bolz, U., Fischer, T. M., Grunewald, S., Scheek, S., Bach, A., Nave, K. A. and Rossner, M. J. (2006). Monitoring regulated protein-protein interactions using split TEV. Nat. Methods 3(12): 985–993.

Xu, M. Q., Paulus, H. and Chong, S. (2000). Fusions to self-splicing inteins for protein purification. Methods Enzymol. 326: 376–418.

Zhong, J., Zhang, H., Stanyon, C. A., Tromp, G. and Finley, R. L., Jr. (2003). A strategy for constructing large protein interaction maps using the yeast two-hybrid system: regulated expression arrays and two-phase mating. Genome Res. 13(12): 2691–2699.

Zhu, H., Bilgin, M., Bangham, R., Hall, D., Casamayor, A., Bertone, P., Lan, N., Jansen, R., Bidlingmaier, S., Houfek, T., Mitchell, T., Miller, P., Dean, R. A., Gerstein, M. and Snyder, M. (2001). Global analysis of protein activities using proteome chips. Science 293(5537): 2101–2105.

Chapter 2
Handling Diverse Protein Interaction Data: Integration, Storage and Retrieval

Benjamin Shoemaker and Anna Panchenko

Abstract In this chapter we review current approaches to store, retrieve and integrate diverse protein interaction data. To incorporate the heterogeneous results of computational predictions and protein interaction experiments, methods of data integration have been widely used which provide efficient presentation, and analysis of interaction data. Among them statistical meta-analysis and supervised machine learning methods are becoming very popular in this respect. While integration methods reduce complexity of system representation, the databases provide efficient storage and retrieval of data. A large variety of interaction databases exist which differ in scope, type and coverage of data as well as query search capabilities. We categorize the databases of protein interactions into comprehensive, specialized, structural and databases developed for network analysis. This gives a rough grouping of resources based on how they might be used. In particular, one might often start with a comprehensive database search and afterwards perform a refined search of the obtained results using a database with a more specific focus.

2.1 Introduction

A protein interactome is a complex and not very well characterized system. The experimental data on protein-protein interactions are obtained under different conditions, for different organisms and provide a wide range of details. For example, Y2H experiments provide the identity of interacting proteins, electron microscopy supplies relative positional information of interacting proteins or protein subunits, and crystallography provides full atomic detail of interaction surfaces. High-throughput experiments can be applied on a genome-wide scale to identify protein interaction partners while more specialized biophysical methods can characterize interactions in terms of the kinetics, dynamics and mechanics of binding processes. In addition to

B. Shoemaker
National Center for Biotechnology Information, National Institutes of Health, Bethesda, USA
e-mail: shoemake@mail.nih.gov

A. Panchenko, T. Przytycka (eds.), *Protein-protein Interactions and Networks*,
DOI: 10.1007/978-1-84800-125-1_2,

direct detection of physical protein interactions, computational methods of protein interaction prediction are becoming more powerful in their ability to identify potential protein interaction partners, specific interacting protein domains and indirect functional associations between proteins.

Integration of heterogeneous experimental and computational interaction data has the ultimate goal of reducing the complexity of a system representation and providing its efficient presentation and analysis. Experimental data collected for a protein interactome contains a lot of false positives while at the same time a large number of relevant interactions are overlooked. Data integration would be the first step to address this problem by emphasizing the bona fide observations supported by multiple data sources and by assigning less confidence to those sporadically observed interactions which can not be considered as reliable and biologically important. Different aspects of data integration have been addressed in various studies but so far the overall problem remains unsolved and most databases treat data sources as complimentary without considering their overlap [1–3].

Diversity and redundancy of experimental and computational interaction data are reflected in a large variety of interaction databases which store the binary protein interactions as well as the higher order interactions in protein complexes. The interaction data are usually obtained by direct submission from experimentalists, by mining literature or by applying computational methods; in some cases the data is verified using automated algorithms or manual curation. Interaction databases differ from one another by the type, size, scale, reliability and coverage of data they contain. They also offer different query capabilities which are provided either by flat files with a textual search engine or by a relational database and more advanced searching capabilities.

In this chapter we give an overview of current approaches to store, retrieve and integrate heterogeneous protein interaction data. In the first part of the chapter we review the two groups of data integration methods: statistical meta-analysis and supervised machine learning methods while the second part of the chapter covers some of the most highly accessible public databases which are categorized in terms of data scope and type.

2.2 Data Integration Methods

2.2.1 Statistical Meta-Analysis

Classical approaches to data integration include various statistical methods of meta-analysis which were developed in an attempt to combine results of several studies and estimate the statistical significance of the combined results [4, 5]. The advantage of these methods is that they do not require any curated training sets which are often unavailable. It was shown that use of conventional statistical procedures in meta-analysis is problematic because it violates the basic assumptions used in conventional statistics, for example, the assumption that the variances associated with every observation are the same or at least do not differ to a large degree. It is not the case when one tries to combine the results of different interaction experiments which have different sample sizes and error rates.

Let us take a set of k independent experiments, each resulting in a set of measurements (ex: gene expression ratios to identify if two genes/proteins interact) which can be characterized by the p-values, $p_1, \ldots, p_k$. Here a p-value estimates the probability to observe a particular data element (gene, protein or their interaction) value purely by chance (the null hypothesis H_{0i} being that the data element/interaction is not affected by the *i-th* experimental perturbation) when, for example, an interaction between two proteins does not exist and/or can not be registered in a particular experiment. The prior reference distribution can be derived from the distribution of non-interacting proteins [6] and measurements with low p-values will be more likely to be true positives. The composite null hypothesis H_0 therefore will hold only if each of the sub-hypotheses $H_{01}, \ldots, H_{0k}$ hold true [4]. Testing of the composite null hypothesis is rather difficult as there are many different alternative hypotheses where at least one of H_{oi} is false. It is illustrated in the p-value space by the complexity of the decision boundary which separates regions which contain mostly true positives with low p-values from those containing mostly true negatives. Intersection and union methods use a predetermined p-value cutoff to detect significant elements resulting in a simple rectangular decision boundary with a large number of missed true positives (intersection method) and a large number of false positives (union method) [7].

One of the most common procedures to combine results of different studies is the Fisher method or inverse chi-square method [4] which uses a product of p-values to combine the effects of k different independent studies. If p-values p_i are distributed uniformly under H_{0i} then $-2 \log p_i$ has a chi-square distribution with two degrees of freedom and consequently the composite p-value has a chi-square distribution with $2k$ degrees of freedom under H_0. Then the null hypothesis will be rejected if

$$P = -2 \sum_{i=1}^{k} w_i \log p_i \geq c,$$

where c can be obtained from the chi-square tables and ω are the weights showing the reliability of different studies. The case with all weights equal to one would correspond to the unweighted Fisher's test. Although it is difficult to select weights *apriori*, the weighted method allows one to emphasize the measurements from more sensitive studies and underrate the less accurate data. Fisher's test can be very informative in the situations where p-values vary among different data sets.

Methods of meta-analysis use different combined test statistics resulting in different decision boundaries and their effectiveness varies depending on the number, size and heterogeneity of the data sets. Recently a new methodology has been developed which was successfully applied to integrate heterogeneous data from different systems biology experiments [7]. According to this methodology the weights for three different integration statistics (including Fisher's statistics P_w) were optimized by maximizing the number of true positives selected at a given significance level (area below decision boundary) in the final set. The maximization was performed by using enhanced simulated annealing and by selecting data elements with combined p-values below the threshold at the final stage. One of the advantages of this method is that it does not make any assumptions about the number of different data sets and can be applied to integrate data from any study. The asymmetric decision boundary produced by this method has been shown to separate very well true positives from true negatives and captures more true positives compared to the unweighted methods.

The method described above was implemented into a free open source software package called Pointillist. In the follow-up study by the same authors, the Pointillist package was applied to integrate 18 different data sources including microarray, protein abundance, chromatin immunoprecipitation experiments, genome-wide data from protein-protein and protein-DNA interaction databases (such as DIP and BIND) and various protein and domain interaction predictions [8]. The data integration procedure resulted in the network for the galactose utilization pathway in yeast where composite p-values were assigned to the nodes and edges of the reconstructed network. The method showed a high sensitivity (99%) for selection of interactions detected by small scale experiments and rejected 93% of protein-protein interactions identified by the single Y2H method and 28% of the interactions detected by multiple Y2H experiments. Similarly to [6] the method reported a 24% false positive rate for interactions predicted by paralog analysis. Based on the analysis of the combined network model the authors were able to predict and explain certain features in the galactose utilization process. Namely, the model suggested that galactose caused the down-regulation of fructose metabolism; an observation which was subsequently verified experimentally.

2.2.2 Supervised Learning Methods

Various methods within the supervised learning framework have been proposed recently to integrate the heterogeneous data to predict protein interactions, to provide confidence levels of predictions and to gain insight into what combinations of features or data sources are the most informative [9–18]. Although these methods use different "gold standard" data sets, feature sets, encodings and learning algorithms for training and testing, it has been shown that the prediction accuracy is improved when several sources of data are combined (for the description of computational methods to predict protein or domain interaction partners, see Chapters 4 and 5). Each protein or protein-protein interaction is encoded as a feature vector where features may represent different data sources (or groups of data sources) on protein-protein interactions such as gene co-expression of two proteins, domain-domain interactions and evidence coming from various experimental and computational methods. The problem of prediction boils down to defining a decision boundary which best separates the data points into interacting and non-interacting protein pairs.

As a result of a comparison of different classifiers, it has been shown that the Random Forest Decision method together with the Support Vector Machine and Logistic Regression methods yield superior performance in the classification tasks [13, 19, 20]. Random Decision Forest classifiers, for example, construct a decision tree which defines a best splitting feature at each node based on how well this feature can discriminate between two classes of interacting and non-interacting proteins. The method has certain advantages as traversing along the tree provides not only the confidence of the prediction but also the information about the contribution of various features at different levels of the decision making process.

One of the Random Forest Decision methods builds decision trees based on the domain composition of interacting and non-interacting proteins, explores all possible combinations of interacting domains and predicts at the end if a given pair of proteins interact [12]. Each protein pair is represented as a vector of length N, where N is the number of different domain types (features), and each feature can have values 2, 1 or 0 depending on whether a domain is found in both proteins, in one of them or not found in the protein pair respectively. Given a training set of interacting protein pairs taken from the experimental data, the method constructs a decision tree (or many trees) which defines the best splitting feature at each node from a randomly selected feature subspace. The best feature is selected based on a measure of "goodness of fit" which estimates how well this feature can discriminate between two classes of interacting and non-interacting pairs. The method stops growing the tree as soon as all pairs at a given node are well separated into two classes providing a classification for an unknown protein pair.

By examining different feature combinations with Random Forest Decision Methods it has been found that the importance of features in correct classification depends on the type of prediction problem (physical protein-protein interactions, co-complex relationships or pathway co-membership), however, gene expression was shown to be the most important feature for all prediction tasks [13]. At the same time in another study using Random Forests together with Logistic Regression it has been shown that MIPS and Gene Ontology (GO) features are the most informative functional similarity datasets for predicting protein-protein interactions. Random Decision Forest classifiers based on the MIPS and GO features alone can yield very accurate predictions [20] (see Chapter 5 for more details).

Bayesian classification methods have proven to be very powerful in the analysis of diverse and noisy interaction data. Similar to Random Forest Decision methods, they provide probabilistic scores for all potential protein interactions; and interactions with the highest confidence levels comprise the predicted protein interaction network. A naïve Bayes classifier, for example, was successfully applied to construct a probabilistic structure for the entire yeast and human interactomes [10, 21]. After assessing relationships between different features used in the first study (ranging from gene co-expression to phylogenetic profile prediction methods) no strong inter-dependency between the features was found [22]. This can be explained either by the fact that different data types encode different types of information or by the incompleteness or biases of different data sources. In the second interactome study, a particular focus was put on the network emerging from cancer genomics data to identify several interaction subnetworks activated in cancer [21].

The large number of information sources on protein-protein interactions makes it particularly hard to integrate this heterogeneous data and extract biological meaning from it. Effective data integration methodologies for protein-protein interactions and systems biology in general should try to accommodate diverse data types from different experimental and computational methods and at the same time, should be able to handle high error rates, missing data and systematic biases in the datasets. Fortunately, recent studies concluded that integrating the information from various experimental and computational studies allows one to predict protein-protein interactions with higher accuracy and to better understand the underlying mechanisms of protein recognition.

2.3 Protein and Domain Interaction Databases

Databases storing information on protein-protein interactions are numerous and diverse. Figure 2.1 shows different types of experimental and computational information input to the various interaction databases (lighter-colored, peripheral boxes).

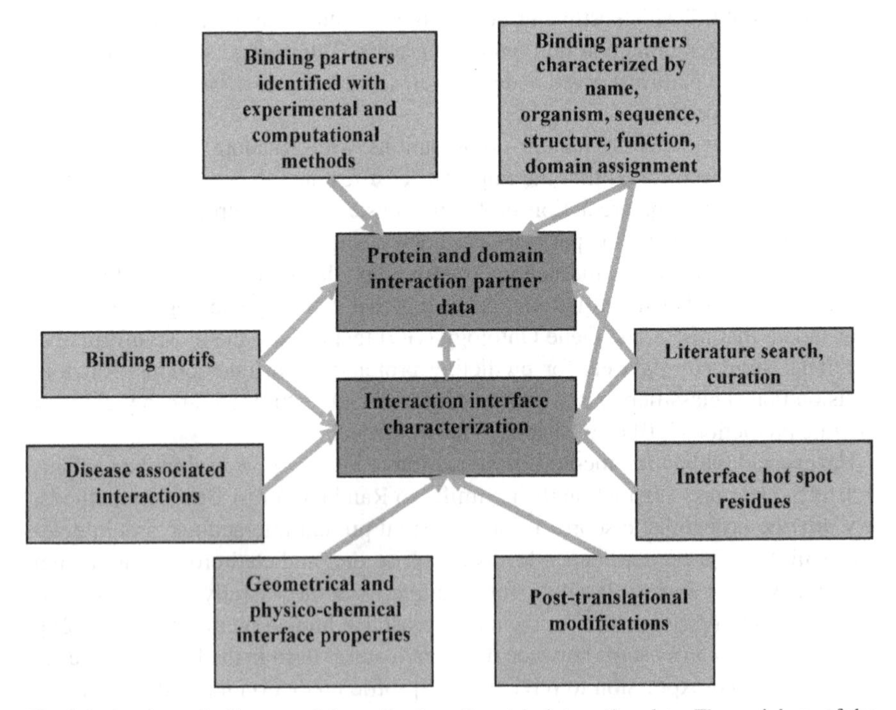

Fig. 2.1 A schematic diagram of the collection of protein interaction data. The periphery of the flow chart shows various sources of experimental data related to protein interactions. The two inner boxes illustrate two general levels of detail collected by the databases

The two inner boxes distinguish the level of detail collected for the protein-protein interactions. In spite of the interaction data diversity, there exist considerable overlaps in the datasets contained in the databases, making it difficult to recommend a single resource for a particular type of information. One attempt to standardize records and reduce the duplication of curation efforts is the formation of the IMEx consortium. Currently there are five active members of the consortium which are noted by the asterisk in Table 2.1. By explicitly exchanging data and coordinating work efforts through IMEx, member databases minimize wasted curator effort. As one example, several member databases are assigned journals to monitor for literature curation and results are shared to the other databases.

The IMEx consortium exchanges all data in the Molecular Interaction (MI) standard developed by the HUPO PSI [32]. This is one of a set of standards setup for data representation in proteomics. With such a diversity of available data types and sources there is a real advantage to the acceptance of such a standard. In particular, many meta-servers and individual researchers who wish to analyze data from multiple sources are greatly assisted by such a database-independent exchange format.

In addition to the many types of data available from databases a variety of displays exist which, although unique for each database, recycle many of the more useful views. Beyond pleasing graphics, the presentation of data helps the user get a quick overview of large amounts of data and make associations that would not otherwise be obvious from simple data downloads. Figure 2.2 shows an example of possible data displays from the STRING database described in the next section. In the top window of the figure there is a summary of the interacting partners to a query protein. This display shows the results both as a graphical network representation and a table including computed relevance scores. For well-connected queries such a display attempts to sort out the more frequent interactors and give an idea of how densely connected the network is.

Here we tried to categorize the databases in terms of the types and scope of data sources (comprehensive versus specialized databases), types of interactions

Table 2.1 Comprehensive databases available for searching and/or downloading data related to protein interactions. Listed are: the name of the database; type of data (high-throughput experimental data (E), structural data (S), manual curation (C) and functional predictions (F)) and the number of interactions. Databases with an asterisk are members of IMEx (http://imex.sourceforge.net)

Database	Type	Number of Entries	URL/FTP
DIP* [23] LiveDIP [24], Prolinks [25]	E,S,F	56,186	http://dip.doe-mbi.ucla.edu, http://prolinks.doe-mbi.ucla.edu/cgi-bin/functionator/pronav/
BOND [26]	E,C,S	83,517	http://bind.ca
STRING [27]	F,E,S	1,513,782 (proteins)	http://string.embl.de
MINT* [28]	E,S,C	103,808	http://mint.bio.uniroma2.it/mint
PubGene	C	—	http://www.pubgene.org
IntAct* [29]	E,C	154,667	http://www.ebi.ac.uk/intact
BioGRID* [30]	E,F,C	198,721	http://www.thebiogrid.org
KEGG BRITE [31]	pathways	9,766 (hierarchies)	http://www.genome.ad.jp/brite

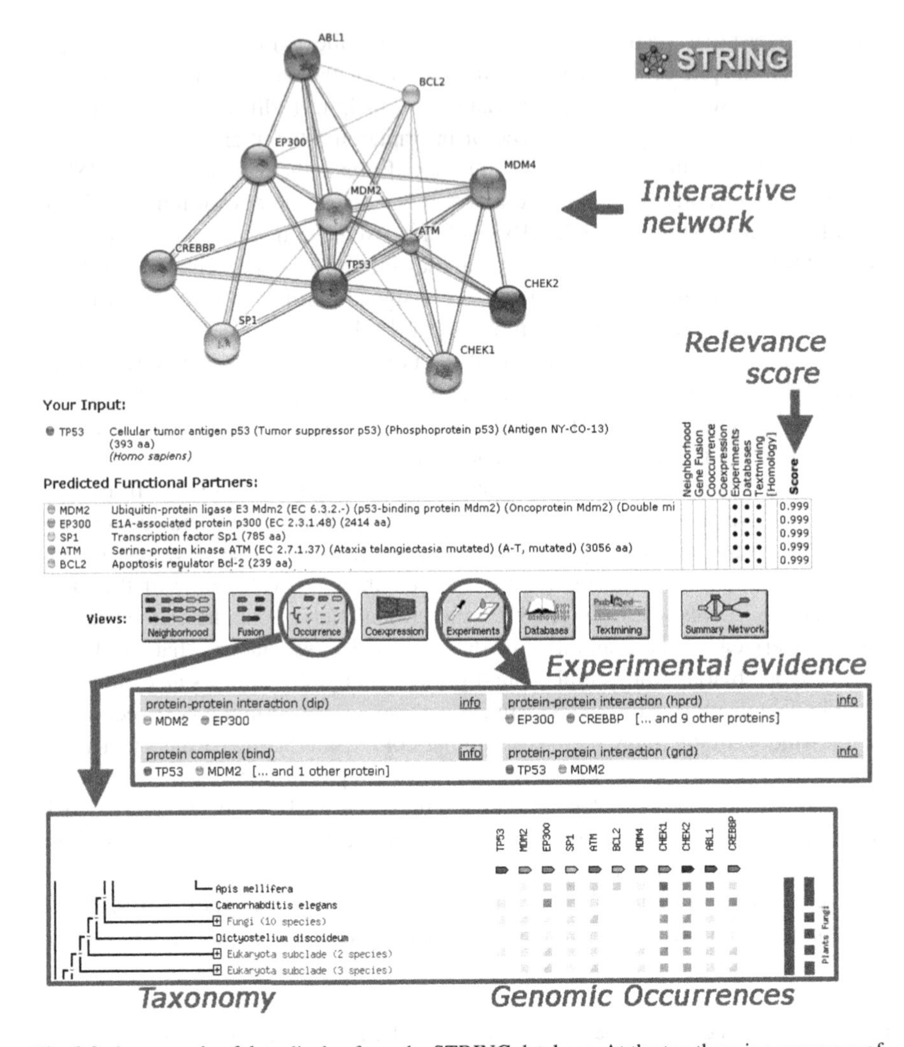

Fig. 2.2 An example of data display from the STRING database. At the top there is a summary of the interacting partners to a query protein. In the middle there are buttons to navigate the displays of relevant supporting evidence. Next there is a table of experimental evidence and graphical display of taxonomy and genome occurrences on the bottom

(physical interactions versus functional associations) or range of details on protein interactions they provide (atomic details derived from structures, interface properties, interacting residues). We also differentiate databases based on the units of interactions into protein-protein and domain-domain interaction databases. We list the most useful features in our opinion for each database which can facilitate the decision of which database to use in search and analysis in each particular case. Below we give brief descriptions of representative databases.

2.3.1 Comprehensive Protein Interaction Databases

In this first section we list databases that collect physical interactions and/or functional associations of protein pairs without any explicit restriction on the focus of the data such as by experimental type, organism or protein function. See Table 2.1 for a full list of comprehensive databases.

2.3.1.1 Database of Interacting Proteins (DIP)

DIP is one of the best known repositories for experimentally-determined protein-protein interactions including a subset of interactions which have passed a quality assessment [23]. Sources for this interaction data range from literature and the Protein Databank (PDB), to high-throughput methods like Y2H, protein microarrays, and TAP / MS analysis of protein complexes (see Chapter 1 for more details). The database makes use of several assessment methods to determine the quality of existing interaction data and to check user-specified interaction sets. DIP can also be accessed via a plugin in Cytoscape [33] to view molecular interaction networks and to integrate with other data such as gene expression profiles. In addition, DIP has links to several related databases including LiveDIP and Prolinks. For proteins in a biological interaction, Live DIP records information about their states and any state changes upon binding, such as covalent modifications, conformations or cellular locations [24]. Prolinks employs four methods of functional association: phylogenetic profiles, Rosetta Stone, gene neighbors and gene clusters [25].

2.3.1.2 Biomolecular Object Network Databank (BOND)

BOND, formerly known as BIND, was one of the first databases along with DIP to collect high-throughput experimental datasets along with other data sources including protein complexes from PDB [26,34]. It holds a large variety of experimental interaction data which were curated by an in-house team of curators. Note, however, that the open access version of this database is no longer being updated. In relation to this database, a data specification was developed to handle various types of protein-protein interaction data as well as protein-small molecule interactions and protein-nucleic acid interactions. BOND uses a grammar of unique icons to distinguish functional types of interactions in displays.

2.3.1.3 Search Tool for the Retrieval of Interacting Proteins (STRING)

The STRING database contains a large number of functional associations between pairs of proteins as well as physically-determined protein-protein interactions. This data comes from four sources: genomic context, high-throughput experiments, conserved coexpression, and previous knowledge [27]. Evidence is listed for functional partners of a query and a scoring scheme is used to judge the relative importance of its interactions.

2.3.2 Specialized Interaction Databases

There are several resources which are focused on a particular subset of interactions or on properties of interaction interfaces. While in most cases more limited in scope, they typically offer more detailed information and higher-quality, curated data sets. Below we will highlight a few such specialized databases and list them in Table 2.2.

2.3.2.1 Human Protein Reference Database (HPRD)

The HPRD collects information for human proteins including a large number of interactions and pathways [35]. Protein-protein interactions come from literature by manual curation. They are linked to the protein records which contain a high level of detail. Annotations include post-translational modifications, disease associations via OMIM, subcellular localizations, protein isoforms and domain architectures, all of which could be useful in characterizing the nature of interactions.

2.3.2.2 Munich MPact/MIPS Database

MPact is the common access point to the MIPS Comprehensive Yeast Genome Database (CYGD) containing a set of manually curated protein-protein interactions from the yeast genome [37]. Interactions in this set have been collected by curators from the literature, but the resource also includes yeast high-throughput results maintained separately.

2.3.2.3 Binding Interface Databases (WikiBID and HotSprint)

WikiBID collects detailed interaction information including bond formation and the locations of interface residues crucial for protein binding ("hotspots") [47]. For each protein pair, WikiBID lists amino acids at the interface and their contributions to binding energies obtained from alanine scanning and site-directed mutagenesis experiments. In addition detailed descriptions of proteins, their interactions and contact maps with bond types are displayed. This database is based on data mined from the literature and is currently updated exclusively by user submissions.

The HotSprint database contains the locations of computational hotspot residues which were predicted based on residue conservation and/or solvent accessible surface area difference upon complexation [48].

2.3.2.4 Molecule Pages Database/UCSD-Nature Signaling Gateway

The Molecule Pages database is a collection of current research on signal transduction proteins from a collaboration between the University of California San Diego (UCSD) and the Nature Publishing Group [36]. Some information has been mined from previous literature, however, what is unique here is that for new articles, interaction information is entered into the database by the original authors and peer

Table 2.2 Specialized protein-protein interaction databases

Database	Scope	Number of Entries	URL/FTP
HPRD [35]	Human, curated	37,581	http://www.hprd.org
Molecule Pages [36]	Cell signaling, curated by author	3,700	http://www.signaling-gateway.org
MPact/MIPS [37]	Yeast, curated	15,488, 4,300 curated	http://mips.gsf.de/services/ppi
PIMdb [38]	Drosophila	—	http://proteome.wayne.edu/PIMproject1.html
SPiD [39]	Bacillus subtilis	112	http://genome.jouy.inra.fr/spid
Kinase Pathway [40]	Kinases, automated literature scan	47,000	http://kinasedb.ontology.ims.u-tokyo.ac.jp
Doodle [41]	Oligomerization domains	1,714	http://dimer.tamu.edu/doodle
pSTIING [42]	Inflammation, cell migration, tumourigenesis	65,228 molecular interactions	http://pstiing.licr.org
Cancer Cell Map [43]	Cancer related pathways	10 pathways	http://cancer.cellmap.org
Scansite [44]	domain-motif interactions, phosphorylation sites	62 motifs	http://scansite.mit.edu
HUGEppi [45]	Human, Y2H, large proteins	84	http://www.kazusa.or.jp/huge/ppi
ASEdb [46]	Experimental interface hotspots	2915 hotspots	http://www.asedb.org
WikiBID [47]	Experimental interface hotspots	7000 hotspots	http://tsailab.tamu.edu/wikiBID
HotSprint [48]	Predicted interface hotspots	30,008 interfaces	http://prism.ccbb.ku.edu.tr/hotsprint
SCOWLP [49]	Interface properties	74,906	http://www.scowlp.org
PINT [50]	Thermodynamic parameters derived from structures	1,700	http://www.bioinfodatabase.com/pint
InterDom [51]	Prediction of domain interactions	30,037	http://interdom.lit.org.sg
DIMA [52]	Domain interactions, phylogenetic profiles	—	http://mips.gsf.de/genre/proj/dima/index.html
DOMINE [53]	Prediction of domain interactions	20513	http://domine.utdallas.edu/cgi-bin/Domine

reviewed as part of the publishing process. This submission requirement minimizes errors made even by expert curators in data mining and means that the data can be extensively typed into various informative categories such as states, transitions, functions and protein classes. For example an author might make a short comment describing two proteins with particular modifications interacting with each other and with small molecules in a pathway.

2.3.2.5 InterDom Database

InterDom collects evidence for predicting protein domain interactions from a number of data sources on protein-protein interactions [51]. These sources include PDB, literature, protein interactions stored in DIP and BOND as well as instances of domain fusion. The reliability of domain interactions is scored depending on the number/type of experimental evidence for each interaction.

2.3.2.6 Domain Interaction Map (DIMA) Database

DIMA uses phylogenetic profiling of Pfam domains to create a domain interaction map [52]. Unlike other similar methods the algorithm is able to save time by avoiding the comparison of entire protein sequences and compares the occurrences of domains across genomes to associate them as interactions based on similar patterns of occurrences. The method also incorporates domain-domain contacts from crystal structures via iPfam and works well in particular for domains with moderate information content which have distinct phylogenetic profiles.

2.3.3 Interaction Databases Using Protein Structures

When available, protein structure data provides a detailed characterization of protein complexes and interaction interfaces. In addition domain-domain interactions within individual proteins can be studied to understand and infer features that might be conserved within protein interaction interfaces. Table 2.3 lists related resources and a few are highlighted here.

2.3.3.1 PIBASE Database

PIBASE is a database of structural domain interfaces, physical protein-protein interactions and structural and functional properties [58], which uses SCOP and CATH domain definitions. Structural comparisons of interfaces are made for the same domain pair within one structure to remove redundancy. The database also combines physicochemical/functional properties of protein binding sites and has a link to MODBASE [63] containing modeled three-dimensional structures which allows one to predict partners and model putative interacting domain interfaces.

Table 2.3 Interaction databases using protein structure

Database	Scope	Number of Entries	URL/FTP
ProtCom [54]	Protein complexes, homology modeling	1,770	http://www.ces.clemson.edu/compbio/ProtCom
3did [55], Interprets[56]	Domain interactions, homology modeling	3,304	http://3did.embl.de
PQS [57]	Quaternary structures	48,568	http://pqs.ebi.ac.uk
Pibase [58], ModBase [59]	Domain interactions, homology modeling	2,387	http://alto.compbio.ucsf.edu/pibase
CBM [60]	Conserved binding modes	2,784	ftp://ftp.ncbi.nlm.nih.gov/pub/cbm
SCOPPI [61]	Structural classification of interfaces	3,358	http://www.scoppi.org
iPfam [62]	Domain interactions	3,019	http://www.sanger.ac.uk/Software/Pfam/iPfam

2.3.3.2 3did Database

3did contains Pfam domain interactions from protein structure data [55]. Views are available for a given domain type or structure for greater detail. For some of the domains, dot plots of structural comparisons show the variance of the interactions between domain pairs. GO-based functional annotations and yeast interactions are also present in the database.

InterPreTS is a web-based service to predict protein-protein interactions based on sequence homology of query proteins to complex structures in a database of interacting domains associated with the 3did database [56].

2.3.3.3 Conserved Binding Mode (CBM) Database

The Conserved Binding Mode (CBM) database collects domain-domain interactions from all PDB structures grouped by geometry into conserved interaction modes for each pair of domain families [60]. Structural alignments are used to infer CBMs from different members of interacting domain families docking in the same way. By searching for domain interactions with recurring structural themes there is a greater chance to find biologically relevant rather than spurious, crystal packing interactions. CBMs highlight the commonalities and variation of a domain pair's interactions from all structural examples.

Currently under development at the NCBI is an interaction tracking database designed to facilitate discovery of interactions related to a query protein or protein domain using the large, inter-connected Entrez databases. All types of physical interactions from protein structure data form the basis of the database, including protein complexes and interactions between protein domains, nucleic acids and small molecules. Sequence homology and structural comparisons are both used to infer interaction partners. Such a scheme allows a user to get a quick overview of possible functional annotations, example structures to view and a wealth of linked information for further details. The strengths of this database lie in its ranking and categorization of interaction types and the scope and update schedule of the underlying data sets.

2.3.3.4 iPfam Database

iPfam displays the interactions of Pfam domains that make contacts in PDB structures [62]. The system is integrated into the Pfam website and allows for interactive browsing of all Pfam-Pfam domain interactions detected on PDB structures at the family and individual structure levels. To illustrate interacting domains, structure, sequence and node displays are shown. In the sequence display details of interacting residue identities are included.

2.3.4 Interaction Network Analysis and Visualization

Table 2.4 lists some of the resources available on the web for additional assistance in visualizing and understanding the topology of interaction networks. Some of these are meta-servers which pull interaction data from multiple host databases and others allow the user to upload their own networks. The network visualization tools can be especially useful for identifying the key elements among the enormous number of connections of protein hub nodes. Here we describe one example of a network analysis tool.

2.3.4.1 PathBLAST

PathBLAST allows a user to compare interaction networks across genomes to identify conserved elements, either complexes or entire pathways [66,71,72]. It currently

Table 2.4 Interaction network analysis and visualization

Tool	Description	Number of Entries	URL/FTP
PIMRider [64]	visualization	3,000	http://pim.hybrigenics.com
Ulysses [65]	interolog analysis	32,930	http://www.cisreg.ca/ulysses
PathBLAST [66]	network alignment	–	http://www.pathblast.org
Cytoscape [33]	visualization & data integration	–	http://www.cytoscape.org
APID [67]	visualization	215,646	http://bioinfow.dep.usal.es/apid
tYNA [68]	network analysis	–	http://networks.gersteinlab.org/tyna
PIANA [69]	data integration, network analysis	–	http://sbi.imim.es/piana
VisANT [70]	visualization	405,301	http://visant.bu.edu

uses data from the DIP database for its computations. First, sequence alignments are made between proteins of the networks from two organisms and their similarity and possible orthology is established. High-scoring network alignments are then constructed from similar proteins which are connected in the same order in two networks. PathBlast can also perform query searches to extract all protein interaction pathways that align with a pathway query of a target organism. There are many different network alignment tools which will be reviewed in Chapter 9.

2.3.5 Conclusion: A Case Study

In conclusion we would like to give an example of one case study of the P53 tumor antigen protein and its protein interactions retrieved by different databases. The tumor antigen p53 protein is a transcription factor regulating many processes of the cell cycle, where it plays a key role in many anti-cancer mechanisms and can respond to DNA damage by inducing cell cycle arrest or apoptosis [73]. Its role as a tumor suppressor has been studied for the last twenty years with the specific focus on its interactions with DNA and numerous protein partners. Many p53 interactions have been identified and studied in detail by experimental techniques and deposited in various databases. We examined the interaction partners for human p53 retrieved from some of the databases listed in Tables 2.1–2.3.

We found that although different databases resulted in rather diverse lists of potentially interacting partners (ranging from 50 to 200 proteins depending on the database), all databases consistently identified the important interaction partners such as MDM2, protein kinases, ubiquitin and the p53-binding protein. Some databases like MINT and STRING provided reliability scores assigned to each potential interaction while offering an excellent viewer of a protein interaction network associated with the p53 protein. At the same time, DIP and BIOGRID supplied detailed information about the experimental techniques and interactions verified by computational methods (in the case of DIP). In addition, specialized databases such as HPRD and SCANSITE provided information on p53 associated diseases, post-translational modifications, predicted phosphorylation and other binding sites. Structure-based databases gave the details of geometrical and physico-chemical properties of interaction interfaces. For example, PiBase and 3did showed specific interacting domains in structures of p53 and its complexes with other proteins. The CBM database found four conserved binding modes of interactions between p53 and other domains/proteins. One of the binding modes involved the interaction with the BRCT domain, and three others represented homo-dimer interactions, which showed conservation between human and mouse species. Finally, several critical residues contributing to thermodynamics and kinetics of binding of p53 to its partners were identified using WikiBID and HotSprint databases.

Acknowledgments This research was supported by the Intramural Research Program of the National Library of Medicine at the National Institutes of Health of the US Department of health and Human Services.

References

1. Birkland A, Yona G: BIOZON: a system for unification, management and analysis of heterogeneous biological data. *BMC Bioinformatics* 2006, **7**:70.
2. Joyce AR, Palsson BO: The model organism as a system: integrating 'omics' data sets. *Nat Rev Mol Cell Biol* 2006, 7(3):198–210.
3. Lacroix Z, Raschid L, Eckman BA: Techniques for optimization of queries on integrated biological resources. *J Bioinform Comput Biol* 2004, **2**(2):375–411.
4. Hedges LV, Olkin I: Statistical methods for meta-analysis: Academic Press; 1985.
5. Hunter JE, Schmidt FL: 'Methods of Meta-Analysis : Correcting Error and Bias in Research' Sage Publications; 1990.
6. Deane CM, Salwinski L, Xenarios I, Eisenberg D: Protein interactions: two methods for assessment of the reliability of high throughput observations. *Mol Cell Proteomics* 2002, **1**(5):349–356.
7. Hwang D, Rust AG, Ramsey S, Smith JJ, Leslie DM, Weston AD, de Atauri P, Aitchison JD, Hood L, Siegel AF et al.: A data integration methodology for systems biology. *Proc Natl Acad Sci U S A* 2005, **102**(48):17296–17301.
8. Hwang D, Smith JJ, Leslie DM, Weston AD, Rust AG, Ramsey S, de Atauri P, Siegel AF, Bolouri H, Aitchison JD et al.: A data integration methodology for systems biology: experimental verification. *Proc Natl Acad Sci U S A* 2005, **102**(48):17302–17307.
9. Gilchrist MA, Salter LA, Wagner A: A statistical framework for combining and interpreting proteomic datasets. *Bioinformatics* 2004, **20**(5):689–700.
10. Jansen R, Yu H, Greenbaum D, Kluger Y, Krogan NJ, Chung S, Emili A, Snyder M, Greenblatt JF, Gerstein M: A Bayesian networks approach for predicting protein-protein interactions from genomic data. *Science* 2003, **302**(5644):449–453.
11. Qi Y, Klein-Seetharaman J, Bar-Joseph Z: Random forest similarity for protein-protein interaction prediction from multiple sources. *Pac Symp Biocomput* 2005:531–542.
12. Chen XW, Liu M: Prediction of protein-protein interactions using random decision forest framework. *Bioinformatics* 2005, **21**(24):4394–4400.
13. Qi Y, Bar-Joseph Z, Klein-Seetharaman J: Evaluation of different biological data and computational classification methods for use in protein interaction prediction. *Proteins* 2006, **63**(3):490–500.
14. Bader JS, Chaudhuri A, Rothberg JM, Chant J: Gaining confidence in high-throughput protein interaction networks. *Nat Biotechnol* 2004, **22**(1):78–85.
15. Lee I, Date SV, Adai AT, Marcotte EM: A probabilistic functional network of yeast genes. *Science* 2004, **306**(5701):1555–1558.
16. Yamanishi Y, Vert JP, Kanehisa M: Protein network inference from multiple genomic data: a supervised approach. *Bioinformatics* 2004, **20** Suppl 1:I363–I370.
17. Bradford JR, Westhead DR: Improved prediction of protein-protein binding sites using a support vector machines approach. *Bioinformatics* 2005, 21(8):1487–1494.
18. Huttenhower C, Troyanskaya OG: Bayesian data integration: a functional perspective. *Comput Syst Bioinformatics Conf* 2006:341–351.
19. Zhang LV, Wong SL, King OD, Roth FP: Predicting co-complexed protein pairs using genomic and proteomic data integration. *BMC Bioinformatics* 2004, **5**:38.
20. Lin N, Wu B, Jansen R, Gerstein M, Zhao H: Information assessment on predicting protein-protein interactions. *BMC Bioinformatics* 2004, **5**:154.
21. Rhodes DR, Tomlins SA, Varambally S, Mahavisno V, Barrette T, Kalyana-Sundaram S, Ghosh D, Pandey A, Chinnaiyan AM: Probabilistic model of the human protein-protein interaction network. *Nat Biotechnol* 2005, **23**(8):951–959.
22. Lu LJ, Xia Y, Paccanaro A, Yu H, Gerstein M: Assessing the limits of genomic data integration for predicting protein networks. *Genome Res* 2005, **15**(7):945–953.
23. Salwinski L, Miller CS, Smith AJ, Pettit FK, Bowie JU, Eisenberg D: The database of interacting proteins: 2004 update. *Nucleic Acids Res* 2004, **32**(Database issue):D449–451.

24. Duan XJ, Xenarios I, Eisenberg D: Describing biological protein interactions in terms of protein states and state transitions: the LiveDIP database. *Mol Cell Proteomics* 2002, **1**(2): 104–116.
25. Bowers PM, Pellegrini M, Thompson MJ, Fierro J, Yeates TO, Eisenberg D: Prolinks: a database of protein functional linkages derived from coevolution. *Genome Biol* 2004, **5**(5):R35.
26. Alfarano C, Andrade CE, Anthony K, Bahroos N, Bajec M, Bantoft K, Betel D, Bobechko B, Boutilier K, Burgess E et al.: The Biomolecular Interaction Network Database and related tools 2005 update. *Nucleic Acids Res* 2005, **33**(Database issue):D418–424.
27. von Mering C, Jensen LJ, Snel B, Hooper SD, Krupp M, Foglierini M, Jouffre N, Huynen MA, Bork P: STRING: known and predicted protein-protein associations, integrated and transferred across organisms. *Nucleic Acids Res* 2005, **33**(Database issue):D433–437.
28. Zanzoni A, Montecchi-Palazzi L, Quondam M, Ausiello G, Helmer-Citterich M, Cesareni G: MINT: a Molecular INTeraction database. *FEBS Lett* 2002, **513**(1):135–140.
29. Hermjakob H, Montecchi-Palazzi L, Lewington C, Mudali S, Kerrien S, Orchard S, Vingron M, Roechert B, Roepstorff P, Valencia A et al.: IntAct: an open source molecular interaction database. *Nucleic Acids Res* 2004, **32**(Database issue):D452–455.
30. Stark C, Breitkreutz BJ, Reguly T, Boucher L, Breitkreutz A, Tyers M: BioGRID: a general repository for interaction datasets. *Nucleic Acids Res* 2006, **34**(Database issue): D535–539.
31. Kanehisa M, Goto S, Hattori M, Aoki-Kinoshita KF, Itoh M, Kawashima S, Katayama T, Araki M, Hirakawa M: From genomics to chemical genomics: new developments in KEGG. *Nucleic Acids Res* 2006, **34**(Database issue):D354–357.
32. Hermjakob H, Montecchi-Palazzi L, Bader G, Wojcik J, Salwinski L, Ceol A, Moore S, Orchard S, Sarkans U, von Mering C et al.: The HUPO PSI's molecular interaction format—a community standard for the representation of protein interaction data. *Nat Biotechnol* 2004, **22**(2):177–183.
33. Shannon P, Markiel A, Ozier O, Baliga NS, Wang JT, Ramage D, Amin N, Schwikowski B, Ideker T: Cytoscape: a software environment for integrated models of biomolecular interaction networks. *Genome Res* 2003, **13**(11):2498–2504.
34. Bader GD, Hogue CW: BIND–a data specification for storing and describing biomolecular interactions, molecular complexes and pathways. *Bioinformatics* 2000, **16**(5):465–477.
35. Mishra GR, Suresh M, Kumaran K, Kannabiran N, Suresh S, Bala P, Shivakumar K, Anuradha N, Reddy R, Raghavan TM et al.: Human protein reference database–2006 update. *Nucleic Acids Res* 2006, **34**(Database issue):D411–414.
36. Li J, Ning Y, Hedley W, Saunders B, Chen Y, Tindill N, Hannay T, Subramaniam S: The Molecule Pages database. *Nature* 2002, **420**(6916):716–717.
37. Guldener U, Munsterkotter M, Oesterheld M, Pagel P, Ruepp A, Mewes HW, Stumpflen V: MPact: the MIPS protein interaction resource on yeast. *Nucleic Acids Res* 2006, **34**(Database issue):D436–441.
38. Pacifico S, Liu G, Guest S, Parrish JR, Fotouhi F, Finley RL, Jr.: A database and tool, IM Browser, for exploring and integrating emerging gene and protein interaction data for Drosophila. *BMC Bioinformatics* 2006, **7**:195.
39. Hoebeke M, Chiapello H, Noirot P, Bessieres P: SPiD: a subtilis protein interaction database. *Bioinformatics* 2001, **17**(12):1209–1212.
40. Koike A, Kobayashi Y, Takagi T: Kinase pathway database: an integrated protein-kinase and NLP-based protein-interaction resource. *Genome Res* 2003, **13**(6A):1231–1243.
41. Marino-Ramirez L, Minor JL, Reading N, Hu JC: Identification and mapping of self-assembling protein domains encoded by the Escherichia coli K-12 genome by use of lambda repressor fusions. *J Bacteriol* 2004, **186**(5):1311–1319.
42. Ng A, Bursteinas B, Gao Q, Mollison E, Zvelebil M: pSTIING: a 'systems' approach towards integrating signalling pathways, interaction and transcriptional regulatory networks in inflammation and cancer. *Nucleic Acids Res* 2006, **34**(Database issue):D527–534.

43. Mathew JP, Taylor BS, Bader GD, Pyarajan S, Antoniotti M, Chinnaiyan AM, Sander C, Burakoff SJ, Mishra B: From bytes to bedside: data integration and computational biology for translational cancer research. *PLoS Comput Biol* 2007, **3**(2):e12.
44. Obenauer JC, Cantley LC, Yaffe MB: Scansite 2.0: Proteome-wide prediction of cell signaling interactions using short sequence motifs. *Nucleic Acids Res* 2003, **31**(13):3635–3641.
45. Kikuno R, Nagase T, Nakayama M, Koga H, Okazaki N, Nakajima D, Ohara O: HUGE: a database for human KIAA proteins, a 2004 update integrating HUGEppi and ROUGE. *Nucleic Acids Res* 2004, **32**(Database issue):D502–504.
46. Thorn KS, Bogan AA: ASEdb: a database of alanine mutations and their effects on the free energy of binding in protein interactions. *Bioinformatics* 2001, **17**(3):284–285.
47. Fischer TB, Arunachalam KV, Bailey D, Mangual V, Bakhru S, Russo R, Huang D, Paczkowski M, Lalchandani V, Ramachandra C et al.: The binding interface database (BID): a compilation of amino acid hot spots in protein interfaces. *Bioinformatics* 2003, **19**(11): 1453–1454.
48. Keskin O, Ma B, Nussinov R: Hot regions in protein–protein interactions: the organization and contribution of structurally conserved hot spot residues. *J Mol Biol* 2005, **345**(5): 1281-1294.
49. Teyra J, Doms A, Schroeder M, Pisabarro MT: SCOWLP: a web-based database for detailed characterization and visualization of protein interfaces. *BMC Bioinformatics* 2006, **7**:104.
50. Kumar MD, Gromiha MM: PINT: Protein-protein Interactions Thermodynamic Database. *Nucleic Acids Res* 2006, **34**(Database issue):D195–198.
51. Ng SK, Zhang Z, Tan SH, Lin K: InterDom: a database of putative interacting protein domains for validating predicted protein interactions and complexes. *Nucleic Acids Res* 2003, **31**(1):251–254.
52. Pagel P, Oesterheld M, Stumpflen V, Frishman D: The DIMA web resource–exploring the protein domain network. *Bioinformatics* 2006, **22**(8):997–998.
53. Raghavachari B, Tasneem A, Przytycka TM, Jothi R: DOMINE: a database of protein domain interactions. *Nucleic Acids Res* 2007, **36**(Database issue):D656–661.
54. Kundrotas PJ, Alexov E: PROTCOM: searchable database of protein complexes enhanced with domain-domain structures. *Nucleic Acids Res* 2007, **35**(Database issue):D575–579.
55. Stein A, Russell RB, Aloy P: 3did: interacting protein domains of known three-dimensional structure. *Nucleic Acids Res* 2005, **33**(Database issue):D413–417.
56. Aloy P, Russell RB: InterPreTS: protein interaction prediction through tertiary structure. *Bioinformatics* 2003, **19**(1):161–162.
57. Henrick K, Thornton JM: PQS: a protein quaternary structure file server. *Trends Biochem Sci* 1998, **23**(9):358–361.
58. Davis FP, Sali A: PIBASE: a comprehensive database of structurally defined protein interfaces. *Bioinformatics* 2005, **21**(9):1901–1907.
59. Pieper U, Eswar N, Braberg H, Madhusudhan MS, Davis FP, Stuart AC, Mirkovic N, Rossi A, Marti-Renom MA, Fiser A et al.: MODBASE, a database of annotated comparative protein structure models, and associated resources. *Nucleic Acids Res* 2004, **32**(Database issue):D217–222.
60. Shoemaker BA, Panchenko AR, Bryant SH: Finding biologically relevant protein domain interactions: conserved binding mode analysis. *Protein Sci* 2006, **15**(2):352–361.
61. Winter C, Henschel A, Kim WK, Schroeder M: SCOPPI: a structural classification of protein-protein interfaces. *Nucleic Acids Res* 2006, **34**(Database issue):D310–314.
62. Finn RD, Marshall M, Bateman A: iPfam: visualization of protein-protein interactions in PDB at domain and amino acid resolutions. *Bioinformatics* 2005, **21**(3):410–412.
63. Pieper U, Eswar N, Davis FP, Braberg H, Madhusudhan MS, Rossi A, Marti-Renom M, Karchin R, Webb BM, Eramian D et al.: MODBASE: a database of annotated comparative protein structure models and associated resources. *Nucleic Acids Res* 2006, **34**(Database issue):D291–295.

64. Formstecher E, Aresta S, Collura V, Hamburger A, Meil A, Trehin A, Reverdy C, Betin V, Maire S, Brun C et al.: Protein interaction mapping: a Drosophila case study. *Genome Res* 2005, **15**(3):376–384.
65. Kemmer D, Huang Y, Shah SP, Lim J, Brumm J, Yuen MM, Ling J, Xu T, Wasserman WW, Ouellette BF: Ulysses - an application for the projection of molecular interactions across species. *Genome Biol* 2005, **6**(12):R106.
66. Kelley BP, Yuan B, Lewitter F, Sharan R, Stockwell BR, Ideker T: PathBLAST: a tool for alignment of protein interaction networks. *Nucleic Acids Res* 2004, **32**(Web Server issue):W83–88.
67. Prieto C, De Las Rivas J: APID: Agile Protein Interaction DataAnalyzer. *Nucleic Acids Res* 2006, **34**(Web Server issue):W298–302.
68. Yip KY, Yu H, Kim PM, Schultz M, Gerstein M: The tYNA platform for comparative interactomics: a web tool for managing, comparing and mining multiple networks. *Bioinformatics* 2006, **22**(23):2968–2970.
69. Aragues R, Jaeggi D, Oliva B: PIANA: protein interactions and network analysis. *Bioinformatics* 2006, **22**(8):1015–1017.
70. Hu Z, Ng DM, Yamada T, Chen C, Kawashima S, Mellor J, Linghu B, Kanehisa M, Stuart JM, DeLisi C: VisANT 3.0: new modules for pathway visualization, editing, prediction and construction. *Nucleic Acids Res* 2007, **35**(Web Server issue):W625–632.
71. Kelley BP, Sharan R, Karp RM, Sittler T, Root DE, Stockwell BR, Ideker T: Conserved pathways within bacteria and yeast as revealed by global protein network alignment. *Proc Natl Acad Sci U S A* 2003, **100**(20):11394–11399.
72. Sharan R, Suthram S, Kelley RM, Kuhn T, McCuine S, Uetz P, Sittler T, Karp RM, Ideker T: Conserved patterns of protein interaction in multiple species. *Proc Natl Acad Sci U S A* 2005, **102**(6):1974–1979.
73. Yin Y, Tainsky MA, Bischoff FZ, Strong LC, Wahl GM: Wild-type p53 restores cell cycle control and inhibits gene amplification in cells with mutant p53 alleles. *Cell* 1992, **70**(6): 937–948.

Chapter 3
Principles of Protein Recognition and Properties of Protein-protein Interfaces

Ozlem Keskin, Attila Gursoy, and Ruth Nussinov

Abstract In this chapter we address two aspects – the static physical interactions which allow the information transfer for the function to be performed; and the dynamic, i.e. how the information is transmitted between the binding sites in the single protein molecule and in the network. We describe the single protein molecules and their complexes; and the analogy between protein folding and protein binding. Eventually, to fully understand the interactome and how it performs the essential cellular functions, we have to put all together - and hierarchically progress through the system.

3.1 Introduction

To accomplish their functions proteins need to interact with other proteins; consequently, physical interactions between proteins are fundamental to biological processes. Some proteins perform localized functions, relevant only within the context of a particular biological process; others may possess a global, high level role, mediating between distinct biological processes (Valente and Cusick 2006). The knowledge of the interactions between proteins is crucial both for prediction of the type of function, and for the understanding of *how* function is performed. Toward this goal, physical protein–protein interactions have been assembled, classified and mapped, culminating in the "interactome", i.e., the complete set of physical protein–protein interactions in a cell. The interactome is often represented as a network of nodes connected by links, where nodes stand for proteins and links for direct physical interactions between them. The interactome provides the global picture of the inter-connectivities; however, in order to understand how the function is dynamically regulated and carried out, the details of the physical interactions are

O. Keskin
Koc University, Center for Computational Biology and Bioinformatics, and College of Engineering, Rumelifeneri Yolu, 34450 Sariyer Istanbul, Turkey
e-mail: okeskin@ku.edu.tr

A. Panchenko, T. Przytycka (eds.), *Protein-protein Interactions and Networks*,
DOI: 10.1007/978-1-84800-125-1_3,

essential. The availability of these assists in the understanding of the dynamics of the system, the signaling and the transiently changing inter-relationships between its components. Eventually, to fully understand *how* the signals propagate between binding sites in the protein and through the network, it is useful to examine the thermodynamic distributions of the populations of conformations around the native state.

Evoking the concept of energy landscapes and folding funnels which emulates the conformational distributions is helpful both for understanding the folding of the polypeptide chain and for the understanding of the protein function, through intra- and inter-molecular recognition and binding. The more flexible the molecule, the larger is the ensemble of diverse conformers and the lower are the barriers between them. The concept of the folding funnel assists in understanding mechanisms in binding. Funnels with rugged bottoms portray and lead to non-specific molecular associations. On the other hand, smoother single or few minima with high barriers imply rigid binding. From the theoretical standpoint, since the complexity of the energy landscape increases rapidly with the size of the system, thus funnels constructed for binding can be expected to be quite complicated (Tsai et al. 1999).

Below, we first illustrate that protein folding and protein-protein binding are similar events with similar underlying principles. We next describe the interactome from the point of view of the different types of associations and protein-protein interfaces. We describe different interface types and compare the properties of interfaces with single chains; and finally, we put the two aspects together, describing the crucial role of the presence of conformational ensembles of single chains and of complexes in the transfer of information across the network. Recognition of the presence of conformational ensembles is extremely important since above all, *signals are transferred by the populations; not by static structures.*

3.2 Protein Folding and Protein Binding are Similar Events

From what we know today, with the exception of binding of inhibitors, it appears that in biological systems, none of the protein molecules functions through a single binding event; rather, function implies cascading through a series of events. For each such event, the population around the bottom of the corresponding funnel serves as the repertoire of potentially available molecules for the following binding event. As in folding funnels, it is not the conformer with the highest population times that will bind in the following step. Rather, it is the conformer whose structure in the *current* bound stage is most favorable for the next binding event. This is a general phenomenon that holds uniformly (Ma et al. 1999; Tsai et al. 1999) in allostery (Kumar et al. 1999), molecular communication, and signal transduction. Conformers whose population times might have been very low in the folding funnels might be considerably enriched as they go down consecutive funnels.

Such a view derives from the understanding that protein folding and protein binding are similar processes, with the only difference between them being the presence (or absence) of chain connectivity. Binding and folding have frequently been referred to as "inter"- and "intra-molecular recognition", emphasizing the fact that the two processes have much in common. Indeed, the hierarchical folding concept can be understood only in these terms. The problems of protein-protein binding and protein folding have been of focal interest for already many years. In general, investigations of these have been addressed at comprehension of aspects of either the binding or the folding. There has been a distinction between the two: protein folding consists of studies of single polypeptide chains, whereas studies of binding address the binding of at least two chains. Despite this traditionally sharp division between the two, it has long been recognized that the types of interactions responsible for these processes are similar, although their relative contributions to stability differ (Argos 1988; Janin and Chothia 1990; Janin et al. 1988; Jones and Thornton 1996; Wu et al. 1994).

The basic difference between binding and folding is the absence (or, presence) of chain connectivity. Yet, it is well known that cleaving the polypeptide chain to create two molecules usually results in a dimer association having a similar structure as the monomer (Kippen et al. 1994). And, at the same time, utilizing a linker to ligate two separate subunits generally results in a similarly folded monomer (Liang et al. 1993). With an extra connection in the intact chain, the folding process is favorable, both kinetically and energetically, *if* the native structure is unaltered. This advantage is due to the favorable entropy. In terms of stability, chain cleavage is entropically favorable due to the splitting of the polypeptide chain (Cheng et al. 1990). In addition, the removal of the linkage constraint may potentially lead to a more favorable binding orientation of the two separate structural parts. Chain cleavage will be favorable if the interactions at the newly formed two-chain interface compensate for the loss of entropy.

3.3 Types of Protein Interactions and Complexes in the Interactome

Inspection of the physical interaction map reveals that most proteins interact with only a few other proteins while a small number of proteins (hubs) have many interaction partners. Hub proteins and non-hub proteins differ in several respects; however, it is still unclear what differentiates between hubs and non-hubs (Ekman et al. 2006). Since a hub interacts with many proteins, and since the surfaces of proteins cannot reasonably be expected to contain as many distinct binding sites, it reasonably follows that binding sites of hub proteins are re-utilized by several or many different partners. Yet, it is not necessarily the case that all hub binding sites are shared. Hubs may contain two types of binding sites: the first is used to bind a *permanent* partner; the second is a shared binding site, used in a *transient*, regulatory binding capacity, associating and dissociating with its partners. Figure 3.1 displays an example of a hub protein, transcriptional activator GCN5 (Han et al. 2004). The connectivity of the protein is very high as displayed in the top figure. Ribbon diagrams shown in the bottom panel show that the same interface might be used to bind to different peptides and a different interface might be used to form another complex. The

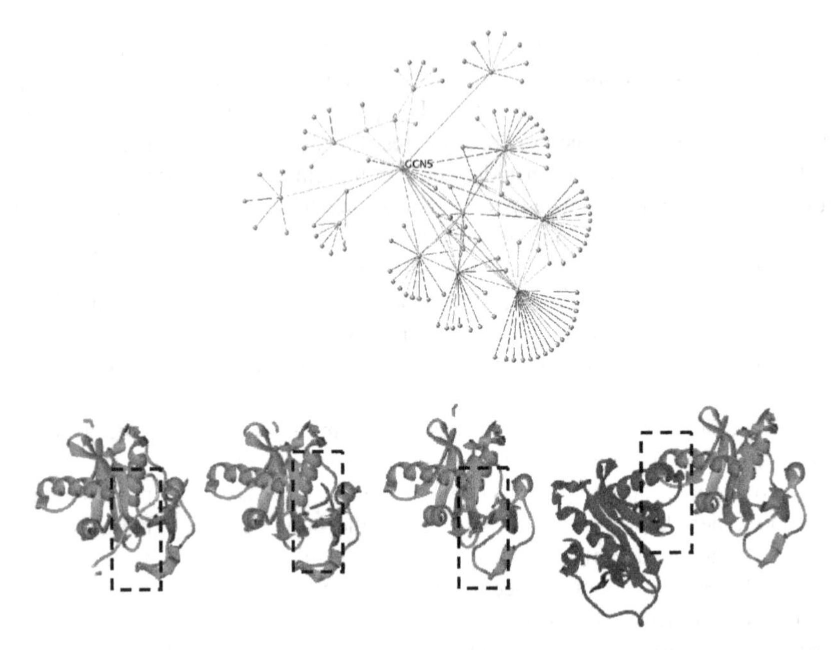

Fig. 3.1 GCN5 histone acetyl transferase is a hub protein. Part (**A**) shows the connectivity of the GCN5 (visualized by VisANT). Part (**B**) shows three complexes with a histone H3 peptide (PDB ID:1pu9), a P53 peptide (PDB ID: 1q2d) and with another histone H3 peptide (PDB ID:1 m1d). These peptides use the same binding site on GCN5. The last ribbon diagram shows the complex between two GCN5 proteins. A different interface is used for dimerization. GCN5 is shown *gray* in the figures whereas the peptide and protein interfaces are boxed

permanently-bound partner may (or, may not) form an *obligatory* interaction. Under such circumstances, the partner is often – though not always – a homo-oligomeric chain, usually a homo-dimer. Homo-dimers are often "two-state" protein-protein complexes, that is, the monomers exist in one of two states: either unfolded in solution, or folded in the complex. In a two-state complex the two protein chains co-fold. Consequently, obligatory homo-oligomeric two-state protein-protein interfaces resemble protein cores. Obligatory associations are found not only in hubs; rather they are a common occurrence in the interactome. On the other hand, shared binding sites imply transient binding to multiple, structurally – and functionally distinct partners. In such cases, the protein chains are also stable on their own, leading to what is generally termed "three-state complexes": here the chain can exist in any of the three states: unfolded in solution; folded on it own; or in a complex. Alternatively, an unstable disordered chain may be permanently and obligatorily bound to another chain, with the shared binding site at a different location than that of the obligatory one. An example of such a case is the elongin B/elongin C complex. On its own, elongin C is unstable. However, it is permanently bound to elongin B, forming a stable two-state complex. A different binding site on elongin C is capable of binding multiple proteins.

Put in these physical interaction terms, the interactome consists of proteins interacting with other proteins through two possible types of associations: obligatory (or, non-obligatory) interactions; and transient (or permanent). Eventually, to be able to perform and coordinate efficiently the complex processes in the cell, both are essential. Obligatory associations are highly specific, perfected by evolution. At the other end of the spectrum, the transient interactions, particularly those involving shared binding sites appear to be much less so, since they have evolved faster, responding to the needs of the organism (Mintseris and Weng 2005). Batada et al. (2006) examined a literature-curated dataset of well-substantiated protein interactions in *Saccharomyces cerevisiae.* They showed that in high quality datasets there is a relatively robust correlation between the rate of evolution and measures of dispensability, or proteins with more interaction partners. Since hub proteins have multiply-utilized binding sites, they do not have a higher density of residues associated with binding. At the same time they undergo rapid turnover and regulation, as observed from high mRNA decay rates and a large number of phosphorylation sites. Thus, with so many proteins to bind to, hubs may also evolve slowly, as some of their interaction sites are constrained in their evolution. On the other hand, their partners may adapt faster, evolving to transiently bind the shared binding site. And by so doing, dynamically inter-connect between modules and drive the newly evolving regulated function.

3.4 Classification into Three Types of Interfaces in the Interactome

To understand the interactome at the structural and functional levels, and its modular organization, the first step is to obtain the physical properties of the protein-protein interactions. Over the years, protein–protein interfaces have been characterized with respect to their structural and physical properties (size, shape, complementarity and packing) and their chemical nature (amino acid composition and conservation, chemical group distributions, hydrophobicity (and hydrophilicity), electrostatic interactions, hydrogen bonding and interactions with water) (Arkin et al. 2003; Nooren and Thornton 2003; Todd et al. 2002). Yet, these properties were largely studied either on all interfaces combined, or separating them into homodimeric and all others. More recently, datasets have been created for permanent *versus* transient complexes. These were largely hand-picked datasets.

In 2002, we have extracted all interfaces between two protein chains obtained from higher complexes of proteins which were available in the protein structural databank (PDB) (Keskin et al. 2004). Interfaces which shared similar architectures were clustered. The interfaces were clustered based on their spatial structural similarities, regardless of the connectivity of their residues on the protein chains. We divided the clusters into three categories: ***Type I*** represents two chain interface clusters with unique functions. Members of these clusters have similar chains and similar interfaces; that is, both the interfaces and the entire protein chains to which the interfaces belong are highly similar and well aligned. Figure 3.2 presents some

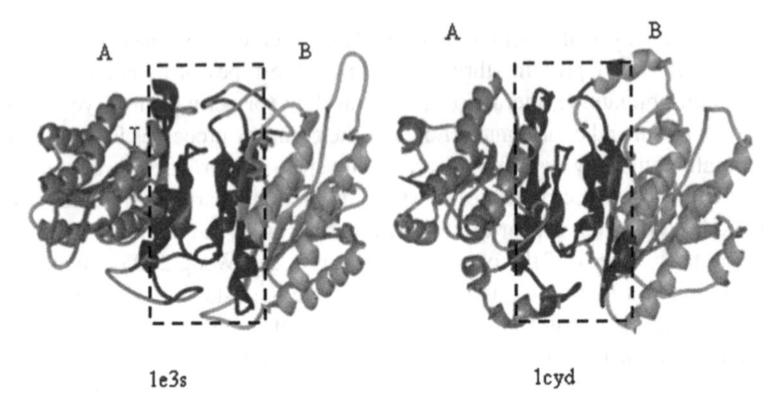

Fig. 3.2 Examples of Type I interfaces. The two proteins are complexes of type II 3-hydroxyacyl-CoA dehydrogenase (HADH II)/amyloid-beta binding alcohol dehydrogenase (PDB ID: 1e3 s) and mouse lung carbonyl reductase (PDB ID: 1cyd). The two different chains of the complexes are shown in different shades of gray. The interface region is shown in black and boxed

examples. ***Type I*** interfaces are very common: in most cases, if the interfaces are similar, the overall protein folds are also similar.

Type II consists of two chain interfaces with multi-functions. The interfaces of members of these clusters are structurally similar; however the global protein folds are different. These similar interfaces, dissimilar protein folds fall into different families (Murzin et al. 1995). However, since they have similar interfaces they are nevertheless members of the same interface clusters. The parent proteins of these interfaces belong to families that have different functions. Hence interface similarity does not ensure global structural similarity. Furthermore, it has been shown previously that globally similar structures may have different functions in proteins, although there is usually a correspondence between fold and function (Moult and Melamud 2000; Nagano et al. 2002; Orengo et al. 1999; Thornton et al. 2000). ***Type II*** interface clusters illustrate that this paradigm can be taken further: similar interfaces do not imply similar functions of the parent proteins from which the interfaces were derived. Figure 3.3 presents examples of ***Type II*** interface clusters. Analysis of ***Type I*** versus ***Type II*** clusters illustrates that (as expected) ***Type I*** is better packed, buries larger total and non-polar accessible surface area, is less planar and has better interface complementarity and more backbone–backbone hydrogen bonds.

The ***Type II*** clusters are extremely interesting: they illustrate that globally different protein structures may associate in similar ways to yield similar motifs. Clearly, in principle, there is a very large number of ways in which monomers can combinatorially assemble. Yet, among these there are preferred interface architectures and these are similar to those observed in monomers. This observation both underscores the view that the number of favorable motifs is limited in nature and highlights the analogy between binding and folding. It is further reminiscent of the combinatorial assembly of protein building blocks in folding. Here, we observe that there are many cases where evolutionarily related proteins have diverged from each other in

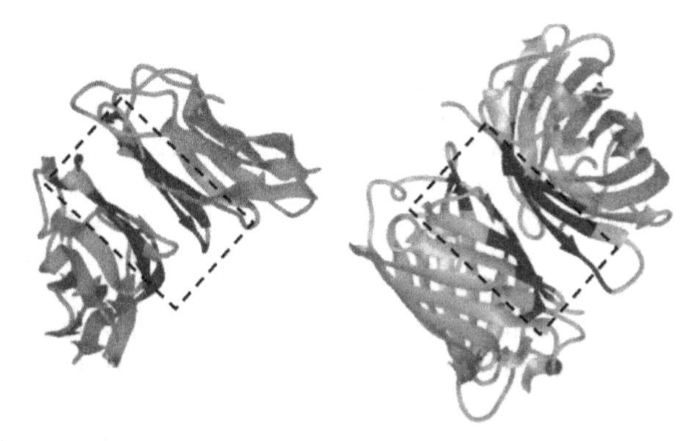

Fig. 3.3 Examples of Type II interfaces. The *left* panel shows the crystal structure of single cohesin domain from the scaffolding protein CIPA (PDB ID: 1aoh). The *right* structure belongs to a luminescent protein (PDB ID: 1b9c). The interface region is shown with a darker shade and boxed

function, yet maintained the interfaces they use to interact with other proteins. The question arises as to whether it is possible to infer from cases such as those in our dataset the time scales of evolutionary divergence.

Type III also represents interfaces with multi-functions, however, unlike ***Type II***, members of these clusters have only one side of their interface aligned. Within a cluster, all proteins whose interfaces belong to the cluster have dissimilar functions. Figure 3.4 provides some examples. Understanding how a given site binds to different binding sites may shed light on the mechanism of protein interactions. If we assume that there is analogy between hub proteins and multi-partner proteins, ***Type III*** cases may assist in understanding hub proteins versus proteins at the network edges. Inspection of the connectivity of our proteins revealed that they have higher numbers of interactions with other proteins (~13) compared with the average connectivity number in yeast interactome (~5) (Grigoriev 2003). Detailed analysis of multi-partner interfaces indicates that proteins that use common interface

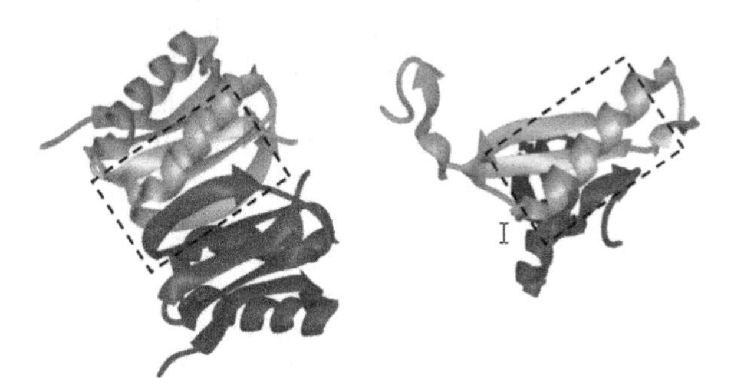

Fig. 3.4 Examples of Type III interfaces. The *left* panel shows the complex structure of dynein light chain 8 (DLC8) (PDB ID: 1f95, Chains A and B). The *right* protein is a 4-oxalocrotonate tautomerase (PDB ID: 1otf, Chains A and E). The chains with similar interfaces have darker colors. The interface region is shown with boxes

motifs to bind to other proteins have smaller interfaces than complexes with specific partners. The average accessible surface area (ASA) of multiprotein interfaces is 1235 Å^2, compared to the 1967 Å^2 ASA of the other types. Most likely, with a large interface it would be more difficult to bind to other, different, complementary sites. We also observed that these multipartner interfaces were not as well packed and organized as other proteins. The geometrical matching was not as optimized, and there were water molecules, allowing more variability in the interactions. We also found that multipartner interfaces preferentially consisted of α helices. Helices appear as the major vehicle through which similar binding sites are able to bind different partners. Helices at multipartner binding sites allow alternate variable ways to achieve favorable binding, depending on the side chain identities. They allow more dynamics in the optimization of the helical associations as compared to extension of β sheets. It will be of interest to examine whether centrally located proteins with multiple proteins binding at the same sites are enriched in α-helical folds as compared to the edge proteins.

3.5 Protein-protein Interfaces and Protein Cores are Similar

Above, we discussed the analogy between the processes of protein folding and protein-protein binding and the types of protein-protein interfaces and complexes. Here we ask to what extent are the protein-protein interfaces actually similar to single chains? The cooperative nature of the folding of two-state protein-protein complexes is the outcome of the hydrophobic effect, similar to its being the driving force in a single-chain folding. In analogy to the protein-folding process, the two-chain, two-state model complex may correspond to the formation of compact, hydrophobic nuclei. On the other hand, the three-state model complex involves binding of already folded monomers, similar to the association of the hydrophobic folding units within a single chain. The similarity between folding entities in protein cores and in two-state protein-protein interfaces, despite the absence of some chain connectivity, indicates that chain linkage does not necessarily affect the native conformation. This further substantiates the notion that tertiary, non-local interactions play a critical role in protein folding. These compact, hydrophobic, two-chain folding units, derived from structurally dissimilar protein-protein interfaces, provide a rich set of data useful in investigations of the role played by chain connectivity and by tertiary interactions in studies of binding and of folding.

Architectural motifs: The general similarity in the forces governing protein folding and protein-protein associations has led us to examine the similarity in the architectural motifs between the interfaces and the monomers. Comparisons between the single-chain protein structural dataset and the interface dataset derived both from all protein-protein complexes in the structural database showed that despite the absence of chain connections, the global features of the architectural motifs, present in monomers, recur in the interfaces, a reflection of the limited set of the folding patterns. However, although similarity has been observed, the details of the architectural motifs vary. In particular, the extent of the similarity correlates with the consideration of how the interface has been formed. Interfaces derived from two-state model complexes where the chains fold cooperatively, display a considerable similarity to architectures in protein cores, as judged by the quality of their geometric superposition. On the other hand, the three-state model interfaces, representing binding of already folded molecules, manifest a larger variability and resemble the monomer architecture only in general outline. The origin of the difference between the monomers and the three-state model interfaces can be understood in terms of the different nature of the folding and the binding that are involved. Whereas in the former all degrees of freedom are available to the backbone to maximize favorable interactions, in rigid body three-state model binding, only six degrees of freedom are allowed. Hence, residue or atom pair-wise potentials derived from protein-protein associations are expected to be less accurate, substantially increasing the number of computationally acceptable alternate binding modes (Finkelstein et al. 1995).

Driving forces: Although the hydrophobic effect plays a dominant role in protein-protein binding, it is not as strong as that observed in the interior of protein monomers. Comparison of interiors of the monomers with those of the interfaces reveals that, in general, the hydrophobic amino acids are more frequent in the interior of the monomers than in the interior of the protein-protein interfaces. On the other hand, a higher proportion of charged and polar residues are buried at the interfaces, suggesting that hydrogen bonds and ion pairs contribute more to the stability of protein binding than to protein folding. Moreover, comparison of the interior of the interfaces to protein surfaces indicates that the interfaces are poorer in polar/charged than the surfaces and are richer in hydrophobic residues. The interior of the interfaces appears to constitute a compromise between the stabilization contributed by the hydrophobic effect on the one hand and avoiding patches on the protein surfaces that are too hydrophobic on the other. Such patches would be unfavorable for the unassociated monomers in solution in three-state complexes. We concluded that, although the types of interactions are similar between protein-protein interfaces and single-chain proteins overall, the contribution of the hydrophobic effect to protein-protein associations is not as strong as to protein folding.

Above, we considered protein chains and protein-protein complexes. Combined they form the interactome. This leads us to consider how are the signals which are generated by the binding of proteins at one binding site propagate to another site. This information transfer is essential for regulation of the biological processes.

3.6 How are Signals Transmitted Through the Network?

Biological systems are networks. To optimally address functional requirements, avoiding waste yet with the right components available at the needed quantities at any given time necessitates orchestration with appropriate switches. Efficiency mandates regulation, which in turn dictates the response accounting for the state of the network and the environment. The response is triggered by the presence or absence of certain interactions with other molecules. Intermolecular interactions are physical binding events: between proteins and proteins, proteins and DNA, proteins and small molecules and drugs; they relate to genetic relationships which govern how genes combine leading to the observed phenotypes. Physical interactions control the switches of cellular machines, sensitive to their quantitative yield versus the dynamically changing needs. Allostery is the vehicle translating and transmitting the effects of these physical interactions.

Under given environmental conditions, allostery regulates the increase or decrease in catalytic activities; the transport of proteins and ligands; and it coordinates enzymatic and signaling pathways. The hallmark of allostery has long been that binding at one site affects the conformation of the other (Daily and Gray 2007; Gunasekaran et al. 2004; Lindsley and Rutter 2006; Swain and Gierasch 2006; Wilson et al. 2007). This occurs through an allosteric inducer, which may be another protein molecule or any other ligand. The inducer interacts with the target protein, and via successive making and breaking of (non-covalent) bonds, the inducer eventually leads to a conformational change at the second site. Yet, crucial to the understanding of allostery is that such events do not create new populations of conformations with altered binding site shapes. Instead, allosteric regulation takes place via the *re-distribution of the existing protein conformational ensembles*. This implies that native protein structures do not consist of a single conformation species; rather, currently there is ample evidence that the native state is a certain distribution of *pre-existing* ensembles of conformational substates some of which already have altered binding site shapes. The allosteric re-distribution increases the relative population of these substates (Fetler et al. 2007). The binding of the allosteric inducer can be viewed as changing the environment or the physical conditions of the target protein; and this change is transmitted, leading to a shift of the distributions of the conformational substates. The two binding sites, that of the allosteric ligand and the one whose shape is altered, may be nearby or far away on the protein surface.

To date, the prevailing view of allostery tends to focus on structure. Yet, since allostery is fundamentally thermodynamic in nature, communication across the protein may be mediated not only by changes in the mean conformation but also by changes in the dynamic fluctuations about the mean conformation. That is, allosteric communication may involve not solely the enthalpic component, which is the key factor responsible for the observed alteration in the binding site shape, but also has an entropic contribution (Hawkins and McLeish 2004; Homans 2005; Popovych et al. 2006; Wand 2001). Currently, there are now clear data illustrating that allostery need not involve a conformational change. Allosteric signals initiating at one site need not culminate in a change in a target site shape. In particular, there are striking examples where it has been convincingly demonstrated that allostery may involve *solely* an entropic component. This appears to close the lid on the central dogma of allostery, stipulating that the inducer binds at one site and induces a conformational change in a second site. This dogma had two components: first, that there are two distinct conformations which do not co-exist; and second, that allostery involves a change of shape. Actually, the term *allostery* comes from *allos*, "other," and *stereos*, "shape, " i.e. a different shape. The first step toward a new view of allostery derived from viewing the native state as consisting of an ensemble, and hence allostery as involving a conformational shift of pre-existing conformations. Yet, the accepted outcome was still a visible change in the binding site shape. Now, current evidence clearly indicates that there may not even be a conformational change. This emphasizes the pre-existence of conformational substates and leads to a new definition of allostery as purely thermodynamic phenomena. A definition in these terms underlines the fact that visual inspection of allosteric and non-allosteric states may not show any differences; and in particular, that the absence of marked shape changes does not imply that allosteric regulation is not involved. The latter has vast implications in recognizing new allosteric switches, and drug target. Thus, allostery is much broader than envisioned by the Monod, Wyman and Changeux "MWC" model. It is an inherent property of protein conformations, embodied in their existence as ensembles; and is governed by thermodynamics, enthalpy and, as we now see in experiment, also – *or even solely* – in entropy.

3.7 Conclusions

This chapter addresses protein-protein interactions and the protein network starting from protein chains and their interactions. We describe the analogy between protein chains and protein-protein complexes, focusing on the similarity between protein folding and protein binding. The hierarchical protein folding model views protein folding as hierarchical binding events of units of the protein chain, with the binding driven by the hydrophobic effect. These units may be building blocks, independently folding hydrophobic units, and domains. This already implies that the associations of these units are similar in nature to protein-protein binding. Protein-protein association is the next hierarchical stage in the binding events, dominated by function. Thus, the interactome is hierarchically built. At the same time, in order to dictate the transient associations and their *dynamically changing state*, signals have to be transmitted through the protein molecule. This signal transfer between two (or more) binding sites is the allosteric effect. Thus, at the lowest level, allostery is a key in regulating the functional state and activity of the interactome.

Acknowledgments This project has been funded in whole or in part with Federal funds from the National Cancer Institute, National Institutes of Health, under contract number NO1-CO-12400. The content of this publication does not necessarily reflect the views or policies of the Department of Health and Human Services, nor does mention of trade names, commercial products, or organizations imply endorsement by the U.S. Government. This research was supported (in part) by the Intramural Research Program of the NIH, National Cancer Institute, Center for Cancer Research.

References

Argos, P. (1988) An investigation of protein subunit and domain interfaces. *Protein Eng* 2, 101–113.

Arkin, M. R., Randal, M., DeLano, W. L., Hyde, J., Luong, T. N., Oslob, J. D., Raphael, D. R., Taylor, L., Wang, J., McDowell, R. S., Wells, J. A., and Braisted, A. C. (2003) Binding of small molecules to an adaptive protein-protein interface. *Proc Natl Acad Sci U S A* 100, 1603–1608.

Batada, N. N., Hurst, L. D., and Tyers, M. (2006) Evolutionary and physiological importance of hub proteins. *PLoS Comput Biol* 2, e88.

Cheng, Y. S., Yin, F. H., Foundling, S., Blomstrom, D., and Kettner, C. A. (1990) Stability and activity of human immunodeficiency virus protease: comparison of the natural dimer with a homologous, single-chain tethered dimer. *Proc Natl Acad Sci U S A* 87, 9660–9664.

Daily, M. D., and Gray, J. J. (2007) Local motions in a benchmark of allosteric proteins. *Proteins* 67, 385-399.

Ekman, D., Light, S., Bjorklund, A. K., and Elofsson, A. (2006) What properties characterize the hub proteins of the protein-protein interaction network of Saccharomyces cerevisiae? *Genome Biol* 7, R45.

Fetler, L., Kantrowitz, E. R., and Vachette, P. (2007) Direct observation in solution of a preexisting structural equilibrium for a mutant of the allosteric aspartate transcarbamoylase. *Proc Natl Acad Sci U S A* 104, 495–500.

Finkelstein, A. V., Badretdinov, A., and Gutin, A. M. (1995) Why do protein architectures have Boltzmann-like statistics? *Proteins* 23, 142–150.

Grigoriev, A. (2003) On the number of protein-protein interactions in the yeast proteome. *Nucleic Acids Res* 31, 4157–4161.

Gunasekaran, K., Ma, B., and Nussinov, R. (2004) Is allostery an intrinsic property of all dynamic proteins? *Proteins* 57, 433–443.

Han, J. D., Bertin, N., Hao, T., Goldberg, D. S., Berriz, G. F., Zhang, L. V., Dupuy, D., Walhout, A. J., Cusick, M. E., Roth, F. P., and Vidal, M. (2004) Evidence for dynamically organized modularity in the yeast protein-protein interaction network. *Nature* 430, 88–93.

Hawkins, R. J., and McLeish, T. C. (2004) Coarse-grained model of entropic allostery. *Phys Rev Lett* 93, 098104.

Homans, S. W. (2005) Probing the binding entropy of ligand-protein interactions by NMR. *Chembiochem* 6, 1585–1591.

Janin, J., and Chothia, C. (1990) The structure of protein-protein recognition sites. *J Biol Chem* 265, 16027-16030.

Janin, J., Miller, S., and Chothia, C. (1988) Surface, subunit interfaces and interior of oligomeric proteins. *J Mol Biol* 204, 155–164.

Jones, S., and Thornton, J. M. (1996) Principles of protein-protein interactions. *Proc Natl Acad Sci U S A* 93, 13–20.

Keskin, O., Tsai, C. J., Wolfson, H., and Nussinov, R. (2004) A new, structurally nonredundant, diverse data set of protein-protein interfaces and its implications. *Protein Sci* 13, 1043–1055.

Kippen, A. D., Sancho, J., and Fersht, A. R. (1994) Folding of barnase in parts. *Biochemistry* 33, 3778–3786.

Kumar, S., Ma, B., Tsai, C. J., Wolfson, H., and Nussinov, R. (1999) Folding funnels and conformational transitions via hinge-bending motions. *Cell Biochem Biophys* 31, 141–164.

Liang, H., Sandberg, W. S., and Terwilliger, T. C. (1993) Genetic fusion of subunits of a dimeric protein substantially enhances its stability and rate of folding. *Proc Natl Acad Sci U S A* 90, 7010–7014.

Lindsley, J. E., and Rutter, J. (2006) Whence cometh the allosterome? *Proc Natl Acad Sci U S A* 103, 10533–10535.

Ma, B., Kumar, S., Tsai, C. J., and Nussinov, R. (1999) Folding funnels and binding mechanisms. *Protein Eng* 12, 713–720.

Mintseris, J., and Weng, Z. (2005) Structure, function, and evolution of transient and obligate protein-protein interactions. *Proc Natl Acad Sci U S A* 102, 10930–10935.

Moult, J., and Melamud, E. (2000) From fold to function. *Curr Opin Struct Biol* 10, 384–389.

Murzin, A. G., Brenner, S. E., Hubbard, T., and Chothia, C. (1995) SCOP: a structural classification of proteins database for the investigation of sequences and structures. *J Mol Biol* 247, 536–540.

Nagano, N., Orengo, C. A., and Thornton, J. M. (2002) One fold with many functions: the evolutionary relationships between TIM barrel families based on their sequences, structures and functions. *J Mol Biol* 321, 741–765.

Nooren, I. M., and Thornton, J. M. (2003) Diversity of protein-protein interactions. *Embo J* 22, 3486–3492.

Orengo, C. A., Todd, A. E., and Thornton, J. M. (1999) From protein structure to function. *Curr Opin Struct Biol* 9, 374–382.

Popovych, N., Sun, S., Ebright, R. H., and Kalodimos, C. G. (2006) Dynamically driven protein allostery. *Nat Struct Mol Biol* 13, 831–838.

Swain, J. F., and Gierasch, L. M. (2006) The changing landscape of protein allostery. *Curr Opin Struct Biol* 16, 102–108.

Thornton, J. M., Todd, A. E., Milburn, D., Borkakoti, N., and Orengo, C. A. (2000) From structure to function: approaches and limitations. *Nat Struct Biol* 7 Suppl, 991–994.

Todd, A. E., Orengo, C. A., and Thornton, J. M. (2002) Sequence and structural differences between enzyme and nonenzyme homologs. *Structure* 10, 1435–1451.

Tsai, C. J., Kumar, S., Ma, B., and Nussinov, R. (1999) Folding funnels, binding funnels, and protein function. *Protein Sci* 8, 1181–1190.

Tsai, C. J., Ma, B., and Nussinov, R. (1999) Folding and binding cascades: shifts in energy landscapes. *Proc Natl Acad Sci U S A* 96, 9970–9972.

Valente, A. X., and Cusick, M. E. (2006) Yeast Protein Interactome topology provides framework for coordinated-functionality. *Nucleic Acids Res* 34, 2812–2819.

Wand, A. J. (2001) Dynamic activation of protein function: a view emerging from NMR spectroscopy. *Nat Struct Biol* 8, 926–931.

Wilson, C. J., Zhan, H., Swint-Kruse, L., and Matthews, K. S. (2007) The lactose repressor system: paradigms for regulation, allosteric behavior and protein folding. *Cell Mol Life Sci* 64, 3–16.

Wu, L. C., Grandori, R., and Carey, J. (1994) Autonomous subdomains in protein folding. *Protein Sci* 3, 369–371.

Chapter 4
Computational Methods to Predict Protein Interaction Partners

Alfonso Valencia and Florencio Pazos

Abstract In the new paradigm for studying biological phenomena represented by Systems Biology, cellular components are not considered in isolation but as forming complex networks of relationships. Protein interaction networks are among the first objects studied from this new point of view. Deciphering the interactome (the whole network of interactions for a given proteome) has been shown to be a very complex task. Computational techniques for detecting protein interactions have become standard tools for dealing with this problem, helping and complementing their experimental counterparts. Most of these techniques use genomic or sequence features intuitively related with protein interactions and are based on "first principles" in the sense that they do not involve training with examples. There are also other computational techniques that use other sources of information (i.e. structural information or even experimental data) or are based on training with examples.

4.1 Introduction

The development of methods to extract protein interaction networks is one of the most exciting developments of this century in Molecular Biology. The combination of high-throughput experimental approaches and computational analysis has provided much information regarding the function of cellular systems in model organisms such as *E coli*, *Yeast*, *C elegans*, *Drosophila* and humans. Such techniques provide an interesting source of biological information on individual components and complexes, and are the basis of systems biology reconstruction and simulation studies.

Even though there are still important limitations in the approaches currently available to explore large interaction spaces, including time, space, modifications

A. Valencia
Structural Computational Biology Programme, Spanish National Cancer Research Centre (CNIO), Madrid, Spain
e-mail: valencia@cnio.es

A. Panchenko, T. Przytycka (eds.), *Protein-protein Interactions and Networks*,
DOI: 10.1007/978-1-84800-125-1_4, © Springer-Verlag London Limited 2008

and dynamics, establishing ambitious initiatives such as the human interactome project (Ideker and Valencia 2006) are positive efforts toward advancing our knowledge and understanding in this area. On the computational side, much effort is being focused on developing protein interaction prediction methods, including those that predict interactions based on sequence similarity and on the corresponding calibration of the possibility of predicting interactions based on similarity levels. Predictions are also being generated with learning systems that are trained with the features of known sets of interactions. Here, we will review the computational methods for predicting protein interactions.

Most of these methods are based on the idea that the process of evolution has left interaction-related traces on the corresponding sequences, protein families and genomic organizations. The simplest model assumes that these signals are the result of the co-evolution of the interacting proteins due to their functional collaboration. In such cases, there is evidence of symmetry in the gene-trees corresponding to the sequences of the interacting proteins, and their genes are maintained in related positions in the corresponding genomes. A number of arguments have been put forward in favor and against this model, stimulating an interesting discussion on the evolutionary model for components of interacting networks. This is reflected in the review we present below of the more interesting developments in the methods to predict interactions, including current results and their limitations.

4.2 Computational Methods vs. Experimental Techniques

4.2.1 Interplay Between Experimental and Computational Methods

Computational and experimental methods to identify interacting proteins are not always completely independent and they may complement each other at many different levels. As in any other large scale high-throughput experiment, a computational processing step is essential to obtain biological information from the raw results, in a much clearer way than in classical low-throughput experiments. In this sense computational biology becomes an intrinsic part of global interactome studies (Bu et al. 2003; Fraser et al. 2002; Han et al. 2004; Jeong et al. 2001; Kelley et al. 2003; Lee et al. 2004; Qin et al. 2003; Wuchty et al. 2003; Yeger-Lotem and Margalit 2003) (and Chapters 1, 7 and 8). Needless to say that the representation, storage and management of such large amounts of data has also required specific computational tools and databases to be developed (Gomez et al. 2005) (and Chapter 2).

Computational approaches have been also used to guide these high-throughput experiments: instead of blindly trying all possible pairs of proteins, a guided selection of baits is performed based on the information obtained in previous pull-downs, hence resulting in higher efficiency and lower cost (Lappe and Holm 2004).

Computational techniques are also intrinsic to the approaches based on the combination of heterogeneous evidences of interaction. Experimental results (either direct interactions or indirect evidences such as mRNA co-expression) can be combined with the ones obtained from purely in-silico methods obtaining in this way highly reliable interactomes (Jansen et al. 2003).

4.2.2 Performance Comparison

In contrast to most bioinformatics tools, the performance achieved by in-silico methods to predict interaction partners is comparable to some of the high-throughput experimental approaches (von Mering et al. 2002). This is particularly true for the first generation high-throughput methods that have high degrees of error when assessed in terms of individual pairs (Aloy and Russell 2002b; Legrain et al. 2001; von Mering et al. 2002). Such wide margins are reflected by the errors in the interactomes obtained by these approaches when evaluated against gold-standard sets of interactions derived from low-throughput experiments. Indeed, this accuracy was estimated to be as low as $\sim$10% in the first high-throughput yeast-two-hybrid experiments (von Mering et al. 2002). This large rate of errors is also probably responsible for the low degree of agreement between similar experiments: i.e. the intersection between the three sets of interacting pairs detected in three independent Yeast Two Hybrid experiments was only of 6 pairs of yeast proteins (Uetz and Finley 2005). Current high-throughput experimental techniques also tend to offer poor coverage and in many cases the methodology has intrinsic limitations that make it possible to test only a fraction of all possible pairs of proteins (Uetz and Finley 2005). Further limitations of the currently used experimental techniques include the tendency to preferentially detect interactions between highly expressed proteins, or between proteins belonging to some cellular compartments in detriment of others (von Mering et al. 2002). These limitations obviously do not affect the in-silico methods which are cheaper and faster than their experimental counterparts.

Obviously, computational methods have their own drawbacks and limitations. For example most of them have difficulties in distinguishing physical from functional interactions. Thus, needless to say that only through the combination of computational (von Mering et al. 2002), or computational and experimental methods (Jansen et al. 2003), are the most interesting results produced in terms of the detection of protein interactions.

4.3 Computational Methods Based on Sequence and Genomic Information

Most of the in-silico methods for predicting interaction partners are based on simple sequence and genome features intuitively related with functional relations between the corresponding proteins (Fig. 4.1). The general rational behind these approaches is that the functional or structural interactions between proteins have potentially modeled their sequences, and/or the organization of the corresponding genes, to

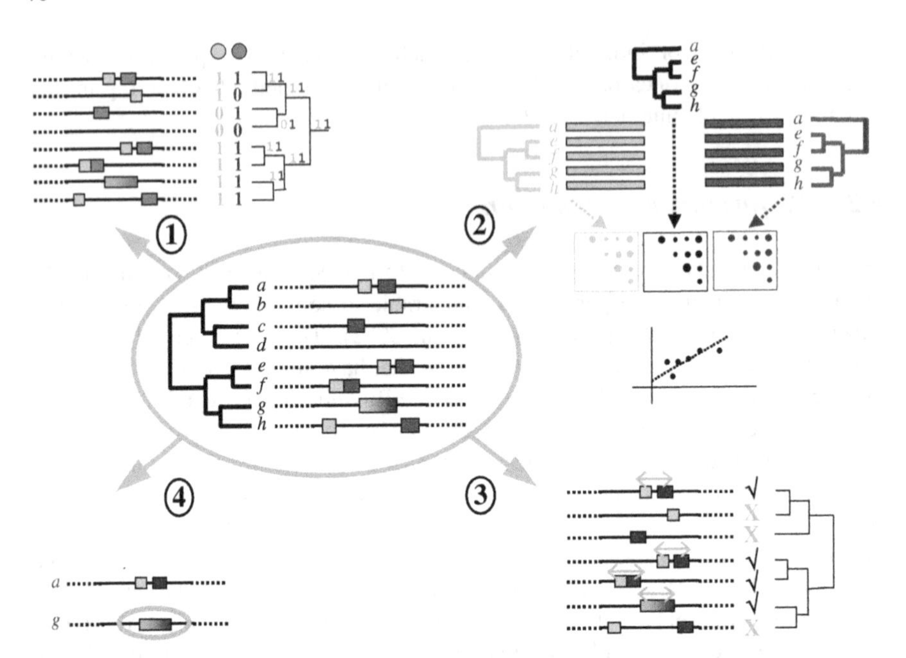

Fig. 4.1 Schemes of the main methods for predicting interacting partners based on sequence and genome information. These methods can be considered to be based on "first principles" since they do not use training and/or extrapolate from known examples. The central part shows the genomic information available for two proteins (*light-gray* and *dark-gray*) for which we want to assess a possible interaction. This information includes the sequences of these proteins and the genome position of the corresponding genes in a number of organisms. (1) Phylogenetic profiling: the presence ("1") or absence ("0") of the two proteins in the different genomes is coded and proteins with similar profiles are predicted to interact. A model of gene gain/loss can be included, together with the phylogeny, to asses the significance of the observed gene profiles. (2) Similarity of phylogenetic trees (*mirrortree*): distance matrices are extracted from the phylogenetic trees of the two proteins built from their sequences in the organisms where both are present. The "background" phylogenetic signal due to the underlying speciation events (which tends to increase the similarity between the trees) may eventually be corrected in different ways, for example by extracting the corresponding distances from the organisms phylogenetic tree. The similarity between the trees is indirectly evaluated as the correlation between these distance matrices. (3) Gene neighboring: if the two proteins are situated nearby in a number of genomes with sufficient phylogenetic distance, they possibly fall under similar transcriptional control and hence, are functionally or physically related. (4) "Rosetta stone" method: if the two proteins are fused in one or more organisms we can say that they are related

better fulfill their potential functions in the corresponding organisms. The following methods exploit this general idea in one way or another.

4.3.1 Phylogenetic Profiling

The "phylogenetic profile" of a given protein family reflects the presence/absence of that family in a set of organisms and as such, it represents the species distribution of the protein family. Pairs of mutually dependent proteins tend to have similar phylogenetic profiles and as such, the two proteins tend to be present in the same subset of organisms and absent from the complementary set (Gaasterland and Ragan 1998; Marcotte et al. 1999b; Pellegrini et al. 1999). The explanation for this fact is that proteins which need each other to perform a given function will either both be present or both absent from a genome since one can not work without the other. The disappearance of both can be explained by "reductive evolution", whereby the organism (especially bacteria) would get rid of one of the genes if the required partner is no longer present.

In their first versions, phylogenetic profiles were coded as binary vectors with "1" coding for the presence of a given gene and "0" coding for its absence (Fig. 4.1). The similarity between vectors was related to protein interactions or functional relationships (Pellegrini et al. 1999). Later, quantitative information was incorporated by encoding in the positions of the vector the similarity of a protein sequence in a given organism with respect to the corresponding sequence in a reference organism, i.e. as BLAST E-values (Altschul et al. 1997) (Date and Marcotte 2003). In this way, these vectors not only contain information about the presence/absence of the proteins, but also regarding their relative divergence.

As more genomes accumulated in the databases, it became necessary to evaluate the influence of the organisms used when constructing the profiles (number, phylogenetic distribution, etc.), as well as other parameters that influenced the performance of the method, such as E-value cut-off for defining "presence" (Sun et al. 2005; Zheng et al. 2002). In general, the method works better the more organisms evenly distributed in phylogenetic terms that are used. Nevertheless, it was also shown that the relationship between the set of genomes used to build the profiles and the performance in detecting functional relationships depends on the functional class of the proteins, being some sets more suitable for some classes than others (Jothi et al. 2007). Some problems related with the uneven phylogenetic distribution of species can be resolved by incorporating information on the phylogeny of the species involved, together with an evolutionary model of gene gain and loss. In this way, profile similarities not due to functional reasons but rather to the underlying evolutionary process are naturally excluded (Barker et al. 2007; Zhou et al. 2006).

"Phylogenetic profiling" has been widely accepted and it has been demonstrated to be a very versatile technique. Not only are similar profiles informative but so are "anti-correlated" ones (where one proteins is present every time the other is absent, or vice-versa). These anti-correlated profiles have been related to enzyme "displacement" in metabolic pathways (Morett et al. 2003). Furthermore, this technique has recently been widened to study protein triplets, facilitating the search for more complicated distribution patterns (e.g. "protein C is present only if A is absent and B is also absent"). This permits the detection of interesting cases representing biological phenomena beyond binary functional interactions, such as complementation (Bowers et al. 2004).

However, this powerful methodology has two important drawbacks. The first is that it can only use sequences from complete genomes (as only in that situation it is possible to be sure of the absence of a given protein). The second is that it can not be applied to the many essential proteins since these tend to be present in all species, resulting in non-informative profiles (containing "1" in all the positions).

4.3.2 Similarity of Phylogenetic Trees

The fact that interacting or functionally related families of proteins apparently present phylogenetic trees with similar topologies had been shown qualitatively in cases such as insulin and its receptors (Fryxell 1996), or dockerins and cohexins (Pages et al. 1997). Such similarities have subsequently been quantified and the relationship with protein interactions statistically demonstrated in large sets of interacting proteins and protein domains (Goh et al. 2000; Pazos and Valencia 2001). In these two studies, the similarity between the trees was indirectly measured as the correlation between the similarity matrices of the protein families (Fig. 4.1).

The similarity of the trees of interacting protein families can possibly be explained by the similar evolutionary pressure exerted on interacting and functionally related proteins, given that they are involved in the same cellular process, and by the fact that they are forced to co-adapt to each other. Both these factors would result in a coordinated evolutionary history or co-evolution, which in turn is reflected in the similarity of the corresponding trees. It is tempting to consider that this co-adaptation will be reflected in the presence of complementary mutations between the corresponding proteins (Section 4.3.5.1). It is important to bear in mind that the usability of the method, as well as the relationship between tree similarity and the interaction of the corresponding proteins, is independent of the underlying evolutionary hypothesis. Distinguishing whether the observed co-evolution of interacting proteins is due to co-adaptation, to other factors (such as similarity of expression levels (Fraser et al. 2004)) or to a combination of them is not straightforward, with data favoring one hypothesis or another (Hakes et al. 2007; Mintseris and Weng 2005). Nevertheless, the co-adaptation hypothesis has driven interesting discussions on the evolution of protein interactions and has been successfully used as a working hypothesis for driving subsequent improvements of the methods (Pazos et al. 2005).

As in the case of phylogenetic profiling, the number and phylogenetic distribution of the species used to build the trees may affect the performance of this *mirrortree* method. The similarity between two trees is affected by many factors besides the co-adaptation of the two proteins. The main factor is the underlying speciation process, which results in a "background" similarity between any pair of trees regardless of the interaction of the corresponding proteins (and both similar to the species tree, the "tree of life"). This is actually the basis of using proteins as "molecular markers", since their trees are expected to be similar to the organism tree. Recently, it was shown that correcting this background similarity improves the performance of the *mirrortree* method (Pazos et al. 2005; Sato et al. 2005). This correction can be achieved by using the phylogenetic distances between species taken from the standard "tree-of-life" based on an accepted molecular marker such as the 16SrRNA (Pazos et al. 2005; Sato et al. 2005), by averaging the values of the distance matrices or by analyzing the principal components of these matrices

(Sato et al. 2005). This *tol-mirrortree* approach also allows non-standard evolutionary events to be detected, such as horizontal gene transfer (HGT), concomitant with the prediction of interactions (Pazos et al. 2005). For this, the 16SrRNA tree is used not only to correct the protein distances but also to asses whether they follow the standard phylogeny it represents or not. Detecting these cases of HGT is important for evolution based interaction prediction methods because, due to their special evolutionary histories, these proteins do not fulfill some of the assumptions these methods are based on (like vertical inheritance). Indeed, it was shown that excluding these automatically detected HGT cases from the predictions increases the predictive performance (Pazos et al. 2005).

It was recently shown that the similarity of distance matrices between interacting proteins is more evident when it is calculated from the residues forming the actual interaction surfaces, rather than using the full protein sequence (Mintseris and Weng 2005). Moreover, excluding high entropy (non-conserved) regions from the alignments before generating the distance matrices was also recently shown to improve performance (Kann et al. 2007). Instead of calculating the similarity between the trees derived from the whole sequences of the proteins, it can be calculated from trees derived from the individual sequence domains within the proteins. This not only permits the detection of which protein interacts with which, but also which particular domains are responsible for this interaction (Jothi et al. 2006) and Chapter 5).

There is another interesting way in which the relationship between tree similarity and interactions can be used. To calculate the similarity between two trees, a "maping" (correspondence between the leaves of both trees) has to be established in order to determine which distances to compare (Fig. 4.1). In what has been discussed so far, all the trees contain orthologues (members of the same family in different species) and hence, the mapping was implicit. Thus, in this case the mapping is known and we want to determine whether the two families interact or not. The opposite situation arises when we know that the two families interact but the mapping is unknown (i.e. which receptor within one family interacts with which ligand in the other). This is very common for pairs of families of paralogues (members of the same family in the same organism), for which specific interactions between some members have been experimentally determined but others have not (i.e. Ras and Ras effectors). Variations of the *mirrrotree* method have been developed to tackle this problem (Izarzugaza et al. 2006; Jothi et al. 2005; Ramani and Marcotte 2003; Tillier et al. 2006). In these methods, different mappings are explored and the one producing the highest similarity between the distance matrices is reported. Since the exhaustive exploration of all possible mappings is not possible, these methods use a Monte Carlo algorithm to perform a guided non-exhaustive exploration of the space of the solutions. The search space is also reduced by disallowing mappings inconsistent with the structure of the underlying phylogenetic trees (Jothi et al. 2005).

An important drawback of *mirrortree* and related approaches is that they can only be applied to pairs of proteins with orthologues in many common species. Only the leaves of the trees corresponding to species where both proteins are present can be used (Fig. 4.1).

4.3.3 Conservation of Gene Neighboring

One of the simplest genomic features related to protein-protein interactions is the tendency of some related genes to be situated close to one another in the genome and for this closeness to be conserved across distant species (Fig. 4.1). This tendency is due to the fact that such physical genomic association may eventually allow the two genes to be co-transcribed in the same mRNA and hence, they may respond to common transcriptional control. Such a tendency is especially evident in prokaryotic organism where it is related to the concept of the "operon". In eukaryotic organisms, transcription control using operons is not common and consequently, the tendency of functionally related genes to be close in the genome is not so evident. The closeness is more informative when it is conserved across distant species, since in closer species the gene neighborhood tends to be similar due simply to the short divergence time. A prototypic example of gene neighborhood conservation is the Tryptophan operon, whose members lie closeby in a number of phylogenetically distant bacteria (Dandekar et al. 1998; Overbeek et al. 1999).

Although it might seem trivial at first to detect these conserved pairs of closely associated genes, the actual methods involve tuning a number of parameters, like the chromosomal distance between the two genes or the phylogenetic distance between the species (Dandekar et al. 1998; Overbeek et al. 1999). The obvious drawback of this approach is its limitation to the bacterial genomes as a source of information, where the tendency to put together functionally related genes in operons is clear. Thus, this methodology can be applied to eukaryotic proteins only if they have homologues in bacteria.

4.3.4 Gene Fusion

It has been seen that the members of some pairs of functionally related proteins tend to be "fused" in the same polypeptide in a number of organisms, a so-called "Rosetta Stone" protein (Fig. 4.1). One example are the two *E coli* proteins involved in histidine biosynthesis HIS2 and HIS10, which are fused into a single polypeptide (HIS2) in Yeast. Indeed, it has been shown that many metabolic proteins are involved in domain fusion events (Tsoka and Ouzounis 2000).

These fusion events are strong indicators of protein interactions and functional relationships, and they have been used to detect such associations in a number of organisms (Enright et al. 1999; Marcotte et al. 1999a). A hypothesis proposed to explain the appearance of such fusion events states that the effective concentration of a complex would be much higher if the two proteins are fused together than if the two proteins are separated and hence have to rely on random motion to find each other to form the active complex (Marcotte et al. 1999a).

A clear advantage of this method is its reliability, since the fact that two proteins fuse is a strong indication of a functional relationship. On the other hand, it has two main disadvantages: i) Ubiquitous domains such as SH3 can lead an automatic method to report functionally meaningless fusions; ii) There are not so many fusion events, especially in prokaryotic organisms, although the existing ones are very informative.

4.3.5 *Other Methods*

Although the methods described in the previous sections are the main ones currently employed, and they have been followed by many authors, there are a number of other computational techniques for predicting partners based on sequence information.

4.3.5.1 Co-evolving Positions

Co-evolving positions in multiple sequence alignments (positions showing a concerted mutational behavior) have been used to predict interaction partners (Pazos and Valencia 2002). The idea is that interacting protein families would present more co-evolving inter-protein positions than non-interacting ones. A working hypothesis states that these co-evolving positions reflect compensatory mutations (mutations in one partner might be compensated by mutations in the other), as found experimentally in some systems (Mateu and Fersht 1999). But it is important to stress that the use of this method is totally independent of whether or not this hypothesis is true, and it only depends on demonstrating a relationship between co-evolving positions and interactions, no matter what causes such co-evolution. It is also important to stress that co-evolution and co-adaptation between two proteins at the residue level does not necessarily have to occur at the interaction interface (Halperin et al. 2006). Compensation might occur between relatively distant positions via allosteric effects. We say "relatively" because it has been demonstrated that these inter-protein correlation signals, even when not exactly at the interface, tend to be closer than the average since they are useful for selecting the right orientation of the two chains in many cases (Pazos et al. 1997).

4.3.5.2 Training-Based Methods

The methods described so far do not involve training. They do not "learn" from examples of known interacting (and non-interacting) pairs, but work only on the basis of "first principles". There are many methods that do involve training with known examples (Ben-Hur and Noble 2005; Bornberg-Bauer et al. 2005; Chen and Liu 2005; Shen et al. 2007; Sprinzak and Margalit. 2006; Sprinzak and Margalit 2001; Yamanishi et al. 2004).

In general, the input to these methods is a set of characteristics (descriptors or attributes) of the proteins or protein pairs, in some cases including experimental data. A machine learning classifier (SVM, neural network, decision tree, etc.) is then fed with these descriptors for examples of interacting and non-interacting pairs, and it "learns" to distinguish these two classes. For example Sprinzak & Margalit used pairs of sequence signatures extracted from known interactions to predict new ones (Sprinzak and Margalit 2001). The domain composition of the proteins is used in various of these training methods, based on the idea that some combinations of domains are more prone to interact than others (Bornberg-Bauer et al. 2005; Chen and Lin 2005; Sprinzak et al. 2006). Indeed, it was shown that these domain signatures are the descriptors that best contribute to the discrimination between interacting and non-interacting pairs (Sprinzak et al. 2006).

In any classification method, the homogeneity of the classes in terms of the descriptors is crucial for the method to work. In the case of protein interactions, it was shown that the attributes for interacting pairs in stable complexes are different from those for transient interactions and hence it is worth separating these classes for prediction purposes (Sprinzak et al. 2006).

4.3.5.3 Structure-Based Methods

Another class of methods for predicting interaction partners could be defined based on the fact that they use structural information. Given the 3D structure of a complex AB, these methods are able to predict whether homologous proteins of A and B will interact or not. For example, Aloy et al. derived statistical potentials from known interactions and then used them to score the possible interactions between the homologues of the members of a given complex (Aloy and Russell 2002a; Aloy and Russell 2003). Similarly, the FOLD-X software has been used to asses the energetic feasibility of different complexes between members of the Ras family and different families of Ras effectors (Kiel et al. 2007).

4.4 Other Computational Methods Not Based on Sequence or Structural Information

There are other computational methods used to predict interacting partners that do not rely on sequence or genomic features but that rather use other data (frequently experimental data). Although the primary source of information for these methods is experimental, we include them here because the interactions derived from them are a secondary result obtained using computational techniques, since the original experiment was not designed to primarily detect interactions. Some experimental techniques specifically designed to detect interactions have been highlighted in Chapter 1.

Gene co-expression has long been used as an indicator of interactions and of functional relationships (Bhardwaj and Lu 2005; Jansen et al. 2002). The fact that the expression profiles of two proteins (in different conditions or time points) are related is a clear indication of co-regulation. The relationship between co-expression and interaction is evident for permanent complexes, such as the ribosome (Jansen et al. 2002). It has also been shown that the expression levels of interacting proteins co-evolve when compared in different organisms (Fraser et al. 2004), and that this co-evolved expression is more evident within functional modules (Chen and Dokholyan 2006).

Genetic interactions, such as synthetic lethality, are also indicative of physical interactions and functional relationships (Qi et al. 2005; Ye et al. 2005). Array-based adaptation of the original techniques to detect these interactions has made it possible to apply them in a high-throughput way, scanning for genetic interactions in whole genomes (Tong et al. 2001; Tong et al. 2004).

4.5 Discussion and Future Trends

With the development of new experimental and computational technologies, biological networks, and in particular protein-protein interaction networks, are among the most relevant areas of research in which progress has been made this century. On the one hand, the information provided by these approaches is extremely interesting for experimental biology where it provides clues about new interactions, pathways and protein complexes. At the same time, and on the basis of the analysis of biological systems, protein networks are the prototypical subject of study of "Systems Biology". Bioinformatics and Computational Biology underlie all the steps in the study of protein interaction networks, from the design of the experiment to the generation and analysis of the data.

The methods for predicting interaction partners from sequence and genome information have become very popular, particularly those related with co-evolution. These methods are in one way or another based on the fact that interacting or functionally related proteins co-evolve, and that this co-evolution is reflected in the sequence or genomic features of the proteins. I.e. a long process of co-evolution at the residue level (coevolving positions) would be reflected in global similarities of the evolutionary histories (similarity of phylogenetic trees – *mirrortree*). In the limit, such co-evolutionay process would lead not only to the co-adaptation of the sequence features but the existences of the proteins themselves as well, removing one partner when the other is not present (phylogenetic profiles). Moreover, the presence of two related proteins in the same operon (gene neighboring) or their fusion in a single polypeptide (gene fusion) are also indications of their concerted (non-independent) evolution. It is important to bear in mind that co-evolution is not the same as co-adaptation: co-evolution is an observation while co-adaptation is a hypothesis that might explains that observation. The evolutionary basis of these methods represents a general limitation since these methods cannot be applied to heterologous interactions (e.g. antigen-antibody).

One important limitation of these computational methods is that they generally predict both, physical and functional interactions, since these generally leave similar landmarks at the sequence and genomic level. An exception are the methods based on training, since this training can be restricted to one type of interaction or another. Another problem is that many of these methods predict interactions between families of proteins rather than individual proteins, since they use evolutionary information from complete families, such as multiple sequence alignments and phylogenetic profiles.

The computational methods used for predicting interaction partners have been shown to nicely complement their experimental counterparts, for example by increasing the reliability of the resulting interactions. Moreover, these computational methods are mature enough to be used by the community. There are repositories like STRING (http://string.embl.de (von Mering et al. 2003)), where the user can look for interacting partners for proteins based on different genomic features, and to obtain some clues on possible cellular roles. This strategy was termed "context-based function prediction", and it is orthogonal and complementary to the traditional sequence-based function prediction.

Maybe in the future we will see a more deep interplay between computational and experimental techniques. Right now, these methodologies remain independent and their results are combined *a posteriori*. But one can think in methods for deciphering interaction partners which combine both methodologies from the very beginning.

The cellular role of the proteins can only be explained in the context of their interactions with others. This is why computational methods for deciphering these interactions are helping in interpreting the huge amounts of genomic data in functional terms.

Acknowledgments We would like to acknowledge the contribution of the members of the Structural and Computational Biology group (CNIO) and the Computational Systems Biology group (CNB-CSIC), especially David de Juan, for interesting discussions. This work was in part funded by the BIO2006-15318 and PIE 200620I240 projects from the Spanish Ministry for Education and Science, and by the LSHG-CT-2003-503265 and LSHG-CT-2004-503567 EU projects.

References

Aloy, P. and Russell, R. B. (2002a) Interrogating protein interaction networks through structural biology. Proc Natl Acad Sci USA, 99, 5896–5901.

Aloy, P. and Russell, R. B. (2002b) Potential artefacts in protein-interaction networks. FEBS Lett, 530, 253–254.

Aloy, P. and Russell, R. B. (2003) InterPreTS: protein Interaction Prediction through Tertiary Structure. Bioinformatics, 19, 161–162.

Altschul, S. F., Madden, T. L., Schaffer, A. A., Zhang, J., Zhang, Z., Miller, W. and Lipman, D. J. (1997) Gapped BLAST and PSI-BLAST: a new generation of protein database search programs. Nucl Acids Res, 25, 3389–3402.

Barker, D., Meade, A. and Pagel, M. (2007) Constrained models of evolution lead to improved prediction of functional linkage from correlated gain and loss of genes. Bioinformatics, 23, 14–20.

Ben-Hur, A. and Noble, W. S. (2005) Kernel methods for predicting protein-protein interactions. Bioinformatics, 21, i38–46.

Bhardwaj, N. and Lu, H. (2005) Correlation between gene expression profiles and protein-protein interactions within and across genomes. Bioinformatics, 21, 2730–2738.

Bornberg-Bauer, E., Beaussart, F., Kummerfeld, S. K., Teichmann, S. A. and Weiner, J., 3rd. (2005) The evolution of domain arrangements in proteins and interaction networks. Cell Mol Life Sci, 62, 435–445.

Bowers, P. M., Cokus, S. J., Eisenberg, D. and Yeates, T. O. (2004) Use of logic relationships to decipher protein network organization. Science, 306, 2246–2249.

Bu, D., Zhao, Y., Cai, L., Xue, H., Zhu, X., Lu, H., Zhang, J., Sun, S., Ling, L., Zhang, N., Li, G. and Chen, R. (2003) Topological structure analysis of the protein-protein interaction network in budding yeast. Nucleic Acids Res, 31, 2443–2450.

Chen, X. W. and Liu, M. (2005) Prediction of protein-protein interactions using random decision forest framework. Bioinformatics, 21, 4394–4400.

Chen, Y. and Dokholyan, N. V. (2006) The coordinated evolution of yeast proteins is constrained by functional modularity. Trends Genet, 22, 416–419.

Dandekar, T., Snel, B., Huynen, M. and Bork, P. (1998) Conservation of gene order: a fingerprint of proteins that physically interact. Trends Biochem Sci, 23, 324–328.

Date, S. V. and Marcotte, E. M. (2003) Discovery of uncharacterized cellular systems by genome-wide analysis of functional linkages. Nat Biotechnol, 21, 1055–1062.

Enright, A. J., Iliopoulos, I., Kyrpides, N. C. and Ouzounis, C. A. (1999) Protein interaction maps for complete genomes based on gene fusion events. Nature, 402, 86–90.

Fraser, H. B., Hirsh, A. E., Steinmetz, L. M., Scharfe, C. and Feldman, M. W. (2002) Evolutionary rate in the protein interaction network. Science, 296, 750–752.

Fraser, H. B., Hirsh, A. E., Wall, D. P. and Eisen, M. B. (2004) Coevolution of gene expression among interacting proteins. Proc Natl Acad Sci U S A, 101, 9033–9038.

Fryxell, K. J. (1996) The coevolution of gene family trees. Trends Genet, 12, 364–369.

Gaasterland, T. and Ragan, M. A. (1998) Microbial genescapes: phyletic and functional patterns of ORF distribution among prokaryotes. Microb Comp Genomics, 3, 199–217.

Goh, C.-S., Bogan, A. A., Joachimiak, M., Walther, D. and Cohen, F. E. (2000) Co-evolution of Proteins with their Interaction Partners. J Mol Biol, 299, 283–293.

Gomez, M., Alonso-Allende, R., Pazos, F., Graña, O., Juan, D. and Valencia, A. (2005) Accessible Protein Interaction Data for Network Modeling. Structure of the Information and Available Repositories. In Priami, C. (ed.), *Transactions on Computational Systems Biology I: Subseries of Lecture Notes in Computer Science*. Springer-Verlag GmbH, Heidelberg, Vol. 3380/2005, pp. 1–13.

Hakes, L., Lovell, S., Oliver, S. G. and Robertson, D. L. (2007) Specificity in protein interactions and its relationship with sequence diversity and coevolution. Proc Natl Acad Sci U S A, 104, 7999–8004.

Halperin, I., Wolfson, H. and Nussinov, R. (2006) Correlated mutations: advances and limitations. A study on fusion proteins and on the Cohesin-Dockerin families. Proteins, 63, 832–845.

Han, J. D., Bertin, N., Hao, T., Goldberg, D. S., Berriz, G. F., Zhang, L. V., Dupuy, D., Walhout, A. J., Cusick, M. E., Roth, F. P. and Vidal, M. (2004) Evidence for dynamically organized modularity in the yeast protein-protein interaction network. Nature, 430, 88–93. Epub 2004 Jun 2009.

Ideker, T. and Valencia, A. (2006) Bioinformatics in the human interactome project. Bioinformatics, 22, 2973–2974.

Izarzugaza, J. M., Juan, D., Pons, C., Ranea, J. A., Valencia, A. and Pazos, F. (2006) TSEMA: interactive prediction of protein pairings between interacting families. Nucleic Acids Res, 34, W315–319.

Jansen, R., Greenbaum, D. and Gerstein, M. (2002) Relating whole-genome expression data with protein-protein interactions. Genome Res, 12, 37–46.

Jansen, R., Yu, H., Greenbaum, D., Kluger, Y., Krogan, N. J., Chung, S., Emili, A., Snyder, M., Greenblatt, J. F. and Gerstein, M. (2003) A Bayesian networks approach for predicting protein-protein interactions from genomic data. Science, 302, 449–453.

Jeong, H., Mason, S. P., Barabási, A. L. and Oltvai, Z. N. (2001) Lethality and centrality in protein networks. Nature, 411, 41–42.

Jothi, R., Cherukuri, P. F., Tasneem, A. and Przytycka, T. M. (2006) Co-evolutionary analysis of domains in interacting proteins reveals insights into domain-domain interactions mediating protein-protein interactions. J Mol Biol, 362, 861–875.

Jothi, R., Kann, M. G. and Przytycka, T. M. (2005) Predicting protein-protein interaction by searching evolutionary tree automorphism space. Bioinformatics, 21, i241–i250.

Jothi, R., Przytycka, T. M. and Aravind, L. (2007) Discovering functional linkages and uncharacterized cellular pathways using phylogenetic profile comparisons: a comprehensive assessment. BMC Bioinformatics, 8, 173.

Kann, M. G., Jothi, R., Cherukuri, P. F. and Przytycka, T. M. (2007) Predicting protein domain interactions from coevolution of conserved regions. Proteins, 67, 811–820.

Kelley, B. P., Sharan, R., Karp, R. M., Sittler, T., Root, D. E., Stockwell, B. R. and Ideker, T. (2003) Conserved pathways within bacteria and yeast as revealed by global protein network alignment. Proc Natl Acad Sci U S A, 100, 11394–11399.

Kiel, C., Foglierini, M., Kuemmerer, N., Beltrao, P. and Serrano, L. (2007) A Genome-wide Ras-Effector Interaction Network. J Mol Biol, 370, 1020–1032.

Lappe, M. and Holm, L. (2004) Unraveling protein interaction networks with near-optimal efficiency. Nat Biotechnol, 22, 98–103.

Lee, I., Date, S. V., Adai, A. T. and Marcotte, E. M. (2004) A probabilistic functional network of yeast genes. Science, 306, 1555–1558.

Legrain, P., Wojcik, J. and Gauthier, J. M. (2001) Protein-protein interaction maps: a lead towards cellular functions. Trends Genet, 17, 346–352.

Marcotte, E. M., Pellegrini, M., Ho-Leung, N., Rice, D. W., Yeates, T. O. and Eisenberg, D. (1999a) Detecting protein function and protein-protein interactions from genome sequences. Science, 285, 751–753.

Marcotte, E. M., Pellegrini, M., Thompson, M. J., Yeates, T. O. and Eisenberg, D. (1999b) A combined algorithm for genome-wide prediction of protein function. Nature, 402, 83–86.

Mateu, M. G. and Fersht, A. R. (1999) Mutually compensatory mutations during evolution of the tetramerization domain of tumor suppressor p53 lead to impaired hetero-oligomerization. Proc Natl Acad Sci U S A, 96, 3595–3599.

Mintseris, J. and Weng, Z. (2005) Structure, function, and evolution of transient and obligate protein-protein interactions. Proc Natl Acad Sci U S A, 102, 10930–10935.

Morett, E., Korbel, J. O., Rajan, E., Saab-Rincon, G., Olvera, L., Olvera, M., Schmidt, S., Snel, B. and Bork, P. (2003) Systematic discovery of analogous enzymes in thiamin biosynthesis. Nat Biotechnol, 21, 790–795.

Overbeek, R., Fonstein, M., D'Souza, M., Pusch, G. D. and Maltsev, N. (1999) Use of contiguity on the chromosome to predict functional coupling. In Silico Biol, 1, 93–108.

Pages, S., Belaich, A., Belaich, J. P., Morag, E., Lamed, R., Shoham, Y. and Bayer, E. A. (1997) Species-specificity of the cohesin-dockerin interaction between Clostridium thermocellum and Clostridium cellulolyticum: prediction of specificity determinants of the dockerin domain. Proteins, 29, 517–527.

Pazos, F., Helmer-Citterich, M., Ausiello, G. and Valencia, A. (1997) Correlated mutations contain information about protein-protein interaction. J Mol Biol, 271, 511–523.

Pazos, F., Ranea, J. A. G., Juan, D. and Sternberg, M. J. E. (2005) Assessing protein co-evolution in the context of the tree of life assists in the prediction of the interactome. J Mol Biol, 352, 1002–1015.

Pazos, F. and Valencia, A. (2001) Similarity of phylogenetic trees as indicator of protein-protein interaction. Protein Eng, 14, 609–614.

Pazos, F. and Valencia, A. (2002) In silico two-hybrid system for the selection of physically interacting protein pairs. Proteins, 47, 219–227.

Pellegrini, M., Marcotte, E. M., Thompson, M. J., Eisenberg, D. and Yeates, T. O. (1999) Assigning protein functions by comparative genome analysis: Protein pylogenetic profiles. Proc Natl Acad Sci USA, 96, 4285–4288.

Qi, Y., Ye, P. and Bader, J. S. (2005) Genetic Interaction Motif Finding by expectation maximization–a novel statistical model for inferring gene modules from synthetic lethality. BMC Bioinformatics, 6, 288.

Qin, H., Lu, H. H., Wu, W. B. and Li, W. H. (2003) Evolution of the yeast protein interaction network. Proc Natl Acad Sci U S A, 100, 12820–12824.

Ramani, A. K. and Marcotte, E. M. (2003) Exploiting the co-evolution of interacting proteins to discover interaction specificity. J Mol Biol, 327, 273–284.
Sato, T., Yamanishi, Y., Kanehisa, M. and Toh, H. (2005) The inference of protein-protein interactions by co-evolutionary analysis is improved by excluding the information about the phylogenetic relationships. Bioinformatics, 21, 3482–3489.
Shen, J., Zhang, J., Luo, X., Zhu, W., Yu, K., Chen, K., Li, Y. and Jiang, H. (2007) Predicting protein-protein interactions based only on sequences information. Proc Natl Acad Sci U S A, 104, 4337–4341.
Sprinzak, E., Altuvia, Y. and Margalit, H. (2006) Characterization and prediction of protein-protein interactions within and between complexes. Proc Natl Acad Sci U S A, 103, 14718–14723.
Sprinzak, E. and Margalit, H. (2001) Correlated sequence-signatures as markers of protein-protein interactions. J Mol Biol, 311, 681–692.
Sun, J., Xu, J., Liu, Z., Liu, Q., Zhao, A., Shi, T. and Li, Y. (2005) Refined phylogenetic profiles method for predicting protein-protein interactions. Bioinformatics, 21, 3409–3415.
Tillier, E. R., Biro, L., Li, G. and Tillo, D. (2006) Codep: maximizing co-evolutionary interdependencies to discover interacting proteins. Proteins, 63, 822–831.
Tong, A. H., Evangelista, M., Parsons, A. B., Xu, H., Bader, G. D., Page, N., Robinson, M., Raghibizadeh, S., Hogue, C. W., Bussey, H., Andrews, B., Tyers, M. and Boone, C. (2001) Systematic genetic analysis with ordered arrays of yeast deletion mutants. Science, 294, 2364–2368.
Tong, A. H., Lesage, G., Bader, G. D., Ding, H., Xu, H., Xin, X., Young, J., Berriz, G. F., Brost, R. L., Chang, M., Chen, Y., Cheng, X., Chua, G., Friesen, H., Goldberg, D. S., Haynes, J., Humphries, C., He, G., Hussein, S., Ke, L., Krogan, N., Li, Z., Levinson, J. N., Lu, H., Menard, P., Munyana, C., Parsons, A. B., Ryan, O., Tonikian, R., Roberts, T., Sdicu, A. M., Shapiro, J., Sheikh, B., Suter, B., Wong, S. L., Zhang, L. V., Zhu, H., Burd, C. G., Munro, S., Sander, C., Rine, J., Greenblatt, J., Peter, M., Bretscher, A., Bell, G., Roth, F. P., Brown, G. W., Andrews, B., Bussey, H. and Boone, C. (2004) Global mapping of the yeast genetic interaction network. Science, 303, 808–813.
Tsoka, S. and Ouzounis, C. A. (2000) Prediction of protein interactions: metabolic enzymes are frequently involved in gene fusion. Nature Genet, 26, 141–142.
Uetz, P. and Finley, R. L., Jr. (2005) From protein networks to biological systems. FEBS Lett, 579, 1821–1827.
von Mering, C., Huynen, M., Jaeggi, D., Schmidt, S., Bork, P. and Snel, B. (2003) STRING: a database of predicted functional associations between proteins. Nucleic Acids Res, 31, 258–261.
von Mering, C., Krause, R., Snel, B., Cornell, M., Oliver, S. G., Fields, S. and Bork, P. (2002) Comparative assessment of large scale data sets of protein-protein interactions. Nature, 417, 399–403.
Wuchty, S., Oltvai, Z. N. and Barabasi, A. L. (2003) Evolutionary conservation of motif constituents in the yeast protein interaction network. Nat Genet, 35, 176–179.
Yamanishi, Y., Vert, J. P. and Kanehisa, M. (2004) Protein network inference from multiple genomic data: a supervised approach. Bioinformatics, 20, I363–I370.
Ye, P., Peyser, B. D., Pan, X., Boeke, J. D., Spencer, F. A. and Bader, J. S. (2005) Gene function prediction from congruent synthetic lethal interactions in yeast. Mol Syst Biol, 1, 2005.0026.
Yeger-Lotem, E. and Margalit, H. (2003) Detection of regulatory circuits by integrating the cellular networks of protein-protein interactions and transcription regulation. Nucleic Acids Res, 31, 6053–6061.
Zheng, Y., Roberts, R. J. and Kasif, S. (2002) Genomic functional annotation using co-evolution profiles of gene clusters. Genome Biology, 3, research0060.0061-0060.0069.
Zhou, Y., Wang, R., Li, L., Xia, X. and Sun, Z. (2006) Inferring functional linkages between proteins from evolutionary scenarios. J Mol Biol, 359, 1150–1159.

Chapter 5
Protein Interaction Network Based Prediction of Domain-Domain and Domain-Peptide Interactions

Katia S. Guimarães and Teresa M. Przytycka

Abstract Protein-protein interaction networks provide important clues about cell function. However, the picture offered by protein interaction alone is incomplete, because techniques for determining interactions at genome scale lack details as to how they are mediated. Stable protein interactions are thought to be largely mediated by interactions between protein domains while transient interactions occur often between small globular domains and short protein peptides, the so called linear motifs. Recently a number of computational methods to predict interactions between two domains and between a domain and a (possibly modified) peptide have been proposed. In this chapter we review representative computational methods focusing on those that use high throughput protein interaction networks to uncover domain-domain and domain-peptide interactions.

5.1 Introduction

Information that can be extracted from protein–protein interaction networks has a growing impact on molecular biology. It facilitates, for example, prediction of protein function (see Chapter 8) and provides insights into the organization and the evolution of protein interaction networks (Chapters 7 and 9). However protein interaction data lacks details on how these interactions are mediated. Full understanding of interaction details would provide a powerful weapon for studying diseases and for designing drug targets. The knowledge of domain interactions and protein domain composition can also be used for prediction of protein-protein interaction (Lander et al. 2001; Sprinzak and Margalit 2001; Wojcik and Schachter 2001; Deng et al. 2002; Shmulevich et al. 2002; Nguyen and Ho 2006; Singhal and Resat 2007).

There are several levels of detail for describing how protein interactions are mediated: from delineating interacting domains to atomic level description of binding

K.S. Guimarães
Center of Informatics, Federal University of Pernambuco, Recife, Brazil
e-mail: katiaguim@gmail.com

A. Panchenko, T. Przytycka (eds.), *Protein-protein Interactions and Networks*,
DOI: 10.1007/978-1-84800-125-1_5,

sites (Chapters 3 and 6). On the highest level, protein interactions are thought to be largely mediated by interactions between domains or between a domain and a peptide (Pawson and Nash 2003). Isolated interacting domains can usually fold independently and are readily incorporated into larger multi-domain proteins.

Domain interactions are quite versatile. Some domain-domain interactions are ***general*** (also called promiscuous (Riley et al. 2005)) meaning that if one protein contains one of the domains and another protein contains the other domain then the two proteins are highly likely to interact. However, many domain interactions, especially the ones involved in cell regulatory systems are highly ***specific*** where in a specific interaction, domains can interact or not, depending on a broader context, like cycle-dependent expression, localization in the cell, specific amino-acid sequence features, etc. For example, the interaction between Cyclin C and Pkinase is specific, since the corresponding domains are present in a large number of non-interacting protein pairs (Riley et al. 2005). Some domains interact only with other domains, others interact with peptides, but some domains (e.g. PDZ) can interact with a domain or a peptide (Pawson and Nash 2003).

Because of importance of the information on binding details for understanding protein interactions, prediction of interacting domains pairs and domain-peptide interactions receive a significant amount of attention in computational biology research. In this chapter, we discuss representative paradigms which are based on high-throughput protein interaction networks.

5.2 Predicting Domain Interactions from Protein Interaction Networks

Most proteins contain two or more domains (Apic et al. 2001) and a protein interaction typically involves binding between two or more specific domains. Interacting domain pairs are often reused within the interactome of an organism and many of them are evolutionarily conserved from prokaryotes to eukaryotes. The relevance of this observation is even more significant in view of recent reports suggesting that domain interactions among several organisms may be more conserved than the protein interactions themselves (Itzhaki et al. 2006).

In this section, we discuss methods that directly use the interaction network to predict domain-domain interaction. As representative methods, we selected Association, Maximum Likelihood Estimation, Domain Pair Exclusion Analysis, Parsimonious Explanation, and an integrative method. Other approaches that also decipher interacting domains from protein interaction networks include support vector machines (Bock and Gough 2001) (supervised learning methods are reviewed in Chapter 2), probabilistic network modeling (Gomez and Rzhetsky 2002), and lowest p-value method (Nye et al. 2005). Obviously, protein interaction network is by no means the only source of information that can be used to predict interacting domains. For example, the gene fusion method (Marcotte et al. 1999), discussed in Chapter 4, can be applied to detect domain interactions (Ng et al. 2003b). Similarly, Pagel and colleagues constructed a domain interaction map based on phylogenetic profiling (Pagel et al. 2004). More recently, Jothi and colleagues proposed mirror tree based approach (see Chapter 4) to identify interacting domain pairs (Jothi et al. 2006; Kann et al. 2007).

For methods that are based on protein-protein interaction network, some domain-domain interactions are more difficult to discover than others. An obvious limitation is the number of experiments which report interactions between proteins mediated by a given domain pair. Additional difficulty arises when a domain pair occurs predominantly in the context of interacting proteins that have multiple ***potential domain contacts***, that is, domain pairs that can potentially mediate a given interaction. In contrast, an interacting domain pair may have one or more ***witnesses,*** that is, interacting single-domain protein pairs in which one protein contains one interacting domain while the second protein contains the other domain. In other words, a witness to a domain interaction is an interacting protein pair which, under the assumption that protein interaction is mediated by domain interaction, can only be explained by interaction between the given domains. Obviously, if an interacting domain pair has enough witnesses to compensate for unreliability of high throughput protein interaction data, discovering such pair is trivial. To separate the trivial predictions from more difficult ones, Riley and colleagues (Riley et al. 2005) associate with each domain a measure called ***modularity***, which is equal to the average number of domains in proteins containing the given domain. A non-trivial prediction of interacting domain pairs would then involve at least one domain, out of the pair, with modularity above some threshold (in their study 2.0). High modularity, however, does not exclude the possibility that a given domain pair has witnesses, and even an isolated occurrence of a domain in a protein with a large number of domains increases the modularity of the domain significantly, without necessarily making the prediction process more difficult. Therefore, Guimarães and colleagues (Guimarães et al. 2006) adopt a more stringent partition into easy and difficult predictions. A domain-domain interaction is considered to be ***difficult*** to predict (from the underlying protein-protein interaction network) if it does not have witnesses and otherwise it is considered easy.

5.2.1 Association Method

Association methods detect over-represented domain pairs in interacting protein pairs. In particular, the method proposed by Sprinzak and Margalit scores each domain pair by the log ratio of the frequency of occurrences in interacting proteins to the frequency of independent occurrences of those domains (Sprinzak and Margalit 2001). That is, if P_i is the observed frequency of domain i in the interaction network and P_{ij} is the observed frequency of domain pair (i, j) as a potential domain contact in interacting protein pairs, then

$$Association\_Score(i, j) = \log \frac{P_{ij}}{P_i P_j}$$

A similar but more sophisticated score has been used by Ng and colleagues (Ng et al. 2003b) in the construction of the domain interaction database InterDom. In their scoring formula they take into account that interactions between proteins with a smaller number of potential domain contacts provide a stronger evidence for domain interactions than interaction between multi-domain proteins, so the interactions are weighted accordingly. The score is computed as:

$$InterDom\_subScore(i, j) = \frac{\sum_{k=1}^{N} \#ex_k \cdot \frac{1}{n_k} \cdot n_k^{ij}}{\sum_{k=1}^{N} \#ex_k \cdot \frac{1}{n_k} \cdot (2 \cdot P_i \cdot P_j)},$$

where N is the number of edges in the protein-protein interaction network, $\#ex_k$ is the number of distinct experiments in the network detecting protein interaction k, n_k is the number of potential domain contacts in protein interaction k, $n_k{}^{ij}$ is the number of potential domain contacts between pair (i, j) in protein interaction k, and P_i is, as before, the observed frequency of domain i in the proteins of the network. A similarly defined score is computed from protein complexes. The full score for an interaction between domains includes, in addition to the two aforementioned terms, an additive term set to 2.0 if a domain pairs is related by fusion event (see Chapter 4), and 0.0 otherwise.

5.2.2 Maximum Likelihood Estimation (MLE)

The main idea of the Maximum Likelihood Estimation (MLE) approach (Deng et al. 2002) is to estimate, for each domain pair, the probability of interaction between domains so that the likelihood of the interaction network is maximized. An important feature of this method is that it allows that the false positives and false negatives of the high-throughput data that constitutes the protein interaction network be explicitly modelled. Here, protein-protein interactions and domain-domain interactions are treated as random variables denoted by P_{AB} and D_{ij}, respectively. $P_{AB} = 1$ if proteins A and B interact, and $P_{AB} = 0$ otherwise. In a similar manner, $D_{ij} = 1$ if domains i and j interact, and $D_{ij} = 0$ otherwise.

Under the assumption that two proteins A and B interact if and only if at least one of their potential domain contacts (i, j) interacts, the probability of interaction between two proteins A and B is obtained as:

$$\Pr(P_{AB} = 1) = 1.0 - \prod_{D_{ij} \in P_{AB}} (1 - \lambda_{ij}), \tag{5.1}$$

where $\lambda_{ij} = \Pr(D_{ij} = 1)$ denotes the probability that domain i interacts with domain j and $D_{ij} \in P_{AB}$ is the set of potential domain contacts in the protein pair (A, B).

Let the random variable O_{AB} describe the experimental observation of an interaction between proteins A and B; $O_{AB} = 1$ if an interaction between proteins A and B is observed and $O_{AB} = 0$ otherwise. Denoting false positive and negative rates respectively by *fp* and *fn* we have

$$\Pr(O_{AB} = 1) = \Pr(P_{AB} = 1)(1 - fn) + (1 - \Pr(P_{AB} = 1))fp. \tag{5.2}$$

The goal of the MLE method is to estimate parameters λ_{ij} to maximize the likelihood function L given by

$$L = \prod_{(A,B)|O_{AB}=1} \Pr(O_{AB} = 1) \prod_{(A,B)|O_{AB}=0} (1 - \Pr(O_{AB} = 1)). \tag{5.3}$$

Hence, denoting by λ the vector composed of all λ_{ij}, the likelihood L is a function of λ, *fp*, and *fn*. Deng and colleagues estimated *fp* and *fn* to be *fp* = 2.5E-4 and *fn* = 0.80. The values λ_{ij} are computed using expectation maximization (EM) that maximizes L. In each iteration, t, values of ${\lambda_{ij}}^{t-1}$ are used to compute $\Pr(O_{AB} = 1|\lambda^{t-1})$ using equations (5.1) and (5.2), and update the parameters using the following Expectation and Maximization steps:

$$\text{Expectation Step: } E(D_{ij}^{AB}) = \frac{\lambda_{ij}^{t-1}(1 - fn)^{O_{AB}} fn^{(1-O_{AB})}}{\Pr(O_{AB}|\lambda^{t-1}, f_n, f_p)}$$

$$Maximization\ Step : \lambda_{ij}^{t} = \frac{1}{N_{ij}} \sum_{A,B} E(D_{ij}^{AB}),$$

where $E(D_{ij}^{AB})$ is the expectation that domain pair (i, j) negotiates the interaction between proteins A and B, and N_{ij} is the number of protein pairs in the network that have (i, j) as a potential domain pair.

5.2.3 Domain Pair Exclusion Analysis (DPEA)

One limitation of the MLE method is its difficulty in detecting specific domain interactions. Indeed, if the interaction between domains i and j is highly specific then λ_{ij} is likely to be small. It has also difficulty in recovering interacting domains which have high modularity. To overcome these weaknesses, Riley and colleagues proposed an alternative domain interaction prediction method, Domain Pairs Exclusion Analysis (DPEA) (Riley et al. 2005). The underlying idea behind this method is the assumption that the maximum likelihood score of a network is, in some sense, a measure of how well the probabilities assigned to putative domain interactions explain the network. Thus, if domain pair (i, j) indeed mediates some protein-protein interactions, then excluding such domain pair as a possible mediator (by fixing the parameter corresponding to λ_{ij} in the MLE method to zero) should decrease the likelihood of interactions between these proteins. This change is measured by value E_{ij} defined as:

$$E_{ij} = \sum_{\substack{\text{protein pairs (A,B) such that}\\ (i,j)\text{ is a potential domain contact}}} \log \frac{\Pr(O_{AB}=1)}{\Pr(O_{AB}=1|\lambda_{ij}=0)} \tag{5.4}$$

where λ_{ij} is the probability of interaction between domains i and j estimated in a way similar to the one used in the MLE method but without including the reliability of the protein interaction network as a component of the likelihood score. Thus, the numerator is the probability that proteins A and B interact, given that domains i and j might interact. The denominator is the probability that proteins A and B interact, given that domains i and j do not interact (also estimated by the expectation maximization procedure where λ_{ij} is set to zero).

The 3,005 domain pairs with E_{ij} at least 3.0 were considered predicted to interact with high-confidence. The DPEA method was able to recover significantly more modular interactions (Riley et al. 2005) confirmed by iPFAM than the MLE method.

5.2.4 Parsimonious Explanation (PE)

The idea of recovering interacting domains by examining how well the potential domain contacts explain the protein interaction network has been developed further by Guimarães and colleagues (Guimarães et al. 2006). Based on the hypothesis that protein interactions evolved in a most parsimonious way, they proposed the *Parsimonious Explanation* (*PE*) method which finds a smallest weighted set of domain interactions that can explain the protein interaction network. This model is formalized as an optimization problem and solved with a Linear Programming procedure. The variables of the linear program represent the potential domain contacts derived from the protein interaction network, and the constraints are given by each protein-protein interaction (edge) in the given network as described below. Those variables can take real values between 0 and 1. The constraint imposed by a given protein interaction enforces that the values of the variables representing the potential domain pairs of that interaction add up to at least 1.0. The construction is illustrated in Fig. 5.1.

According to the parsimony principle, the objective function aims to minimize the overall sum of the variables x_{ij}. Formally, if PDP is the set of the potential domain pairs found in the protein interaction network, and PPI is the set of protein-protein interactions in the given network, then the linear program is given by:

$$Minimize \sum_{(i,j)\in PDP} x_{ij}$$

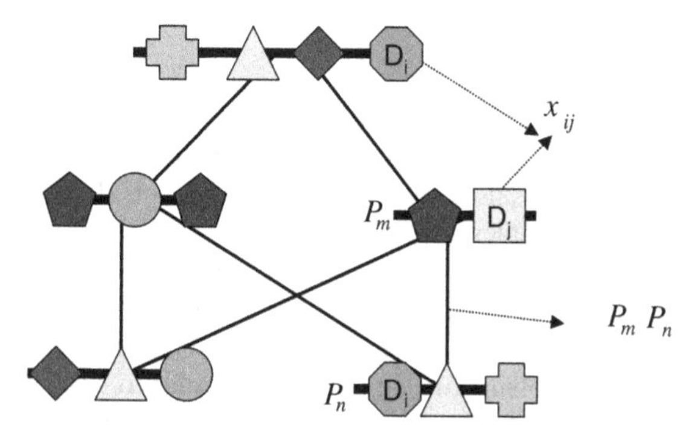

Fig. 5.1 Construction of a Linear Program from a given protein interaction network

$$Subject\ to : \forall(A,B) \in PPI(\sum_{(p,q)\in(A,B)} x_{pq}) \geq 1$$

The value assigned to the variable x_{ij} reflects the contribution of the domain pair (i, j) in explaining the network under the maximum parsimony principle. In the PE method, the false positives of the protein interaction network are modeled by performing a randomization process. In particular, 1000 instances of protein interaction network are constructed which are sub-networks of the input network where each edge is maintained with probability equal to the estimated reliability of the network (in (Guimarães et al. 2006) this value was set to 0.5). For each such randomized network, the corresponding linear program is constructed using the procedure described above, and solved. The reported score, called *LP-score*, of a given domain pair (i, j), is computed by the arithmetic average of the values x_{ij} returned by these 1000 linear programs.

In addition to the LP-score, the PE method offers the so called pw˙score which quantifies the confidence in the LP-score. The pw˙score of a domain pair (i, j) is computed as the minimum of two measures, the p-value of domain pair (i, j), computed from simulations, and a confidence estimation provided by the possible existence of witnesses. The combined witness and p-value score is expressed as:

$$pw\_score = \min(p\_value(i, j), (1 - r)^{w(i,j)})$$

where r is the estimated reliability of the network and $w(i,j)$ is the number of witnesses of domain pair (i, j).

Unlike the previously discussed methods, the Parsimonious Explanation method was able to detect a significant number of difficult interactions confirmed by crystal structures in iPFAM.

5.2.5 Integrative Approaches

With the exception of the scoring function of (Ng et al. 2003a), all methods discussed so far were based exclusively on protein interaction data and protein domain composition. More recently, Lee and colleagues (Lee et al. 2006) proposed a Bayesian approach that complements the protein interaction data with other information about domains; we call their method *Integrative Bayesian* (*IB*).

In the IB method, the expectation of the domain pair interaction is computed separately for each of four organisms, yeast, worm, fruit fly, and humans. The scores for the domain pairs are obtained using a method similar to MLE. The likelihood function is the same as the one used by the MLE method (Deng et al. 2002), however, instead of using $\Pr(D_{ij} = 1)$ directly to score the domain interactions, the IB method scores each domain pair by the expectation of the domain pair interaction given by

$$E(\#D_{ij}) = N_{ij} \cdot \Pr(D_{ij} = 1),$$

where, as before, N_{ij} is the number of protein pairs in the network that have (i, j) as a potential domain contact.

The results obtained for the four networks are considered as four independent pieces of information and used as features in the integrative model. Two additional features considered are the number of times the two domains in the pair appear together, or co-exist, in one protein chain, and the information if the two domains belong to the same biological process as assessed by the Gene Ontology (GO) database (Harris et al. 2004). The scores of all domain pairs with respect to each distinct feature are binned. The likelihood score of a domain pair is computed based on the ratio of domain pairs confirmed by crystal structures to the number of domain pairs not confirmed by crystal structures in the bin containing the score of the given domain pair.

It is interesting to examine how the information which is not obtained based in protein interaction influences the prediction of this method. To elucidate this, Lee and colleagues (Lee et al. 2006) performed a comparison using the domain pairs in iPFAM as true positives, and the remaining domain pairs as true negatives. The results of that comparison are reported in Fig. 4 of (Lee et al. 2006) which shows the relationship between the false positive rate (FP/(FP+TN)) and the sensitivity (TP/(TP+FN)) of the predictions based on different combinations of information. By this evaluation standard, the Gene Ontology terms combined with the domain co-existence gives a better iPFAM pairs recovery than information obtained from the interaction networks using MLE type analysis (see also the discussion in the next section).

Very recently, Wang et al. (2007) introduced a different integrative method, InSite. In addition to the evidences used in the IB method described above they included Prosite (Hulo et al. 2006) motifs treating them in the same way as protein domains. Unlike previous methods, they score domain contacts in the context dependant manner. That is, the score of the same domain pair depends on the protein pair where a given domain pairs makes a potential contact. To obtain such score they use a method similar to the one proposed by Riley et al. (Riley et al. 2005) (see Section 5.2.3 of this chapter, equation (5.4)). However, rather than looking at the effect of disallowing all interactions between a given domain pair, they consider the effect of disallowing single instance of such interaction as possible mediator of a particular protein interaction. This allowed to measure how well given domain interaction explains the given protein interaction rather than its role in predicting all interactions.

5.2.6 Evaluation of Domain-Domain Interaction Prediction Methods

Due to a low coverage of experimentally confirmed domain-domain interactions, evaluation of the accuracy of genome scale methods to predict domain-domain and domain-peptide interactions poses a formidable challenge. One method used to evaluate the quality of predictions is by estimating how accurately one can reconstruct the protein interaction network based on the assigned domain-domain interaction scores (Deng et al. 2002). However the quality of prediction of protein-protein interaction is not necessarily a good measure of correct prediction of domain-domain interaction. While domain pairs that make non-specific interactions are good predictors of protein interactions, the specific domain interactions are not.

An alternative method for assessing predictions was proposed by Nye et al. (Nye et al. 2005). The basic idea is to test if in each pair of interacting proteins, the domain pair with the highest score is correctly predicted as the domain pair mediating the interaction. The test set contains only interacting protein pairs with multiple potential domain contacts and at least one domain pair that is known to interact (e.g. based on the information from the iPFAM database). Guimarães and colleagues (Guimarães et al. 2006) used this method applied to 1780 protein interactions to compare the performances of several domain-domain prediction methods. In that assessment, the Association and the MLE methods achieved a positive predictive value (PPV = TP/(TP+FP)) around 11%, far below the 27% obtained if a potential domain contact had been chosen at random for each protein pair in the set. The DPEA and the PE methods achieved PPV values of 43% and 75%, respectively. That comparison used the Expectation Maximization scores of Riley and colleagues (Riley et al. 2005). Since, unlike the other methods compared, the IB method excludes PFAM-B as possible interacting domains, and its predictions were made based on a different data set, IB was not included in the above comparison. However, in a similar estimation including only 456 protein interactions whose potential domain contacts all have IB score above 0.0, the performance of the IB method is similar to that of PE approach (Guimarães and Przytycka (2008)). The InSite method has been published when this review was virtually completed. It uses a different data set and the scores for the domain pairs were not made available at this time so could not be included in the comparison.

Another method often used to evaluate the quality of domain interaction predictions estimates how well a given method recovers known domain-domain interactions. In this approach, known domain interactions (e.g. domain interactions from iPFAM (Finn et al. 2006), 3DID (Stein et al. 2005) or CBM (Shoemaker et al. 2006)) are considered as true positive and all other domain pairs as true negatives. Under this assumption one can make false positive rate versus true positive rate (or similar) plots. Indeed, if a method is successful, then the corresponding curve should demonstrate a performance clearly better than expected by chance. Among the methods discussed, the highest percentage of iPFAM domains in the top 50 predicted interactions has the InSite method (Wang et al. 2007). However, the number of experimentally confirmed domain-domain interactions is very small relatively to the number of estimated domain-domain interactions. According to a recent study involving E. coli, yeast, worm, fly, and human data, conducted by Itzhaki and colleagues (Itzhaki et al. 2006), the percentage of protein-protein interactions to which high-confidence domain-domain interactions from iPFAM or 3DID could be mapped is no more than 20% for any of the organisms. Therefore, any domain-domain prediction method that undertakes the task of explaining protein interactions through domain-domain interactions is expected to *correctly* recover domain pairs that are not in those high-confidence databases yet. Furthermore, it is expected that domain interactions in iPFAM are highly based towards certain interactions (Guimarães and Przytycka 2008).

Riley et al. bypassed the above problem by selecting a set of true positives among known interacting domain pairs and a set of true negatives (of a similar size) as a set containing domain pairs which belong to interacting protein pairs but do not interact (as confirmed based on available crystal structures of protein complexes). Under this assumption, they tested how many of such true positives and true negatives have been correctly predicted. Using this criterion they estimated that the DPEA method has the specificity of 97% and the sensitivity of 6% (Riley et al. 2005).

Finally, while evaluating domain interaction prediction methods one has to be careful to avoid circularity. Due to a greater interest in some specific domains or functional roles, it is quite possible for some methods to be trained on one type of data and then be evaluated on data that is indirectly related to the one used for training, bringing up opportunity for an artificially inflated performance. Methods that use functional annotation data in particular risk for such circularity (Zhang et al. 2004; Suthram et al. 2006).

5.3 Predicting Domain-Peptide Interactions from Protein Interaction Networks

The methods to predict domain-domain interactions described in the previous section rely on the assumption that protein interactions are mediated by domain-domain interactions. This assumption is well supported for stable protein complexes. However much of the signaling, trafficking, and targeting is mediated by reversible interactions between small globular domains and short protein peptides, the so called *linear motifs*. One of the best studied examples is the SH3 domain which binds to

Proline reach motif PxxP (where x represents any arbitrary amino-acid). A linear motif may, but does not have to be part of a globular domain. In fact, most of such motifs are not (Puntervoll et al. 2003; Neduva et al. 2005). Furthermore, domain-peptide interactions are often very specific, that is homologous domains often bind to different (although related) linear motifs. For example, PxxP is the canonical binding motif for the SH3 domain while a motif for a subclass is often more specific (Toro et al. 2001). One of the first computational problems considered in the context of domain–peptide binding is that of identifying linear motifs that are recognized by a given binding domain (Reiss and Schwikowski 2004; Ferraro et al. 2006; Lehrach et al. 2006). In these approaches experimentally determined SH3 domain-peptide interactions serve as a training set for discovering binding motifs of SH3 domains.

Recently it has been recognized that high throughput interaction networks also provide valuable resource in prediction of protein-peptide interactions. In this section we discuss briefly two methods that take advantage of this information.

5.3.1 Discovering Domain-Peptide Interactions from Protein Interaction Networks

The short length of linear motifs makes their reliable discovery computationally challenging. Recently, several related approaches have been developed that find statistically over-represented motifs in non-homologous sequences with a common property, for example that bind to a certain kinase or phosphatase (Neduva et al. 2005; Davey et al. 2006). We describe here the method of Neduva et al., since this method combines discovery of liner motifs with prediction of direct domain-peptide interaction based on high throughput protein interaction networks.

In the work of Neduva and colleagues (Neduva et al. 2005) the putative linear motifs are identified as sequence fragments observed sufficiently often in protein sequences after removing globular domains (identified as PFAM-A domains), trans-membrane segments, coiled-coils, collagen regions, and signal peptides. Furthermore, homologous sequences were also identified and removed. This preprocessing reduces the probability of detecting motifs shared due to evolutionary relationship or sequence motifs associated with structural motifs such as β-turns. In that approach, all non-overlapping motifs of 3–8 residues are identified using program TEIRESIAS (Rigoutsos and Floratos 1998). Common motifs are required, in particular, to agree perfectly on at least two positions and to occur in at least tree sequences in the set. The neighbors of each protein in the interaction network are examined for occurrences of such common motifs. A common motif observed to be overrepresented among the interacting partners of a given protein is predicted to be the binding motif.

In addition to finding binding motifs of individual proteins, Neduva et al. also searched for the more general binding motifs of homology domains. To do this, they merged sets of binding partners of proteins containing a common domain. Such merged "domain sets" were then analyzed in the same manner as described above for individual proteins.

The analysis of the results obtained with this method is quite revealing. Despite the fact that the data in the protein-protein interaction networks is error-prone, the results were quite accurate, although the number of confidently predicted motifs was relatively small (11 in yeast, 26 in fly, 27 worm, and 112 in human). In all organisms under study, many of the known motifs were missed, as demonstrated by inspection, due to too few sequences with the correct motif to reach significance. The better results for the human network are attributed to the better quality of the data in hand-curated human interactions (Peri et al. 2003) used in the study. The domain set approach was, in some instances, successful in detecting less specific motifs. For example, in the fly network the SH3 motif has been only identified on this level since there was not enough data to detect the more specific binding motifs. The authors have been able to confirm experimentally some of the predicted motifs.

5.3.2 Utilizing Protein Interaction Network in Discovering Phosphorylation Networks

Signal transduction is the primary means by which cells respond to external stimuli such as nutrients, growth factors, and stress. The dynamics of cell signaling pathways is, in large extend, governed by reversible phosphorylation (Krebs and Beavo 1979) of specific substrates performed by protein kinases. Thousands of in vivo phosphorylation sites have been discovered by targeted biochemical studies and, more recently, through spectrometry (Hjerrild et al. 2004). However our understanding of phosphorylation-dependent signaling networks remains incomplete. In particular, despite advances in in-vitro experiments (Ptacek et al. 2005) it is not fully known which protein kinases are responsible for the phosphorylation of many known phosphorylation sites.

There are several computational approaches towards mapping phosphorylation sites to corresponding kinases which are based on identifying consensus sequence motifs recognized by specific kinases (Obenauer et al. 2003; Hjerrild et al. 2004). However, these motifs alone are often insufficient for a unique identification of the kinases responsible for the phosphorylation of the corresponding sites. Specificity of kinase activity is also achieved through cellular localization, cell-cycle specific co-expression, binding to scaffold proteins, etc. Such information, termed by Linding et al. "contextual" (Linding et al. 2007), if available, should also be used to enhance the accuracy of prediction of phosphorylation networks. Along these lines, a recent approach, NetworKIN, combines the motif based and contextual approach into one two-step algorithm.

During the first step on the NetworKIN algorithm, an experimentally determined phosphorylation site is mapped to a protein sequence. Then the protein family that is likely to be responsible for the phosphorylation of the site is predicted based on the consensus motif approach. This is done by applying a neural network machine learning approach to obtain position specific scoring matrices (PSSMs) (Obenauer et al. 2003; Hjerrild et al. 2004) describing biding motifs of all kinase families under study. Once the family (or families) of kinases whose members can potentially phosphorylate a given site is identified, the candidate proteins that could be responsible for the phosphorylation of the site are identified by BLAST search.

In the second stage, the set of candidate kinases in narrowed down using contextual information. The contextual information is obtained from the STRING database (von Mering et al. 2007). This data base integrates information from curated pathways, co-occurrence in abstracts of scientific articles, physical protein interactions, co-expression, and predicted interaction based on genomic context (gene fusion, gene neighborhood, and phylogenetic profiles). All scoring schemas for all evidences were benchmarked and calibrated on signaling and metabolic pathways from KEGG database (Kanehisa et al. 2006) resulting in probabilistic scores for all evidence types. Additionally, association from orthologous protein in other organisms were included using a Bayesian scoring scheme to combine the evidence. The resulting probabilistic association network is used to find kinases that are proximal to the substrate (the protein containing the given phosphorylation site). Namely, for every candidate kinase, the Floyd-Warshall algorithm (Cormen et al. 2001) is used to compute the most likely path connecting this kinase to the substrate. A set of kinases with the best paths are predicted as responsible for the phosphorylation.

The work of Linding et al. demonstrated that the network-based contextual information has a tremendous impact on the prediction accuracy of phosphorylation. The authors estimated that 80% of the predictive power of their approach comes from the contextual information.

5.4 Conclusions and Future Directions

In this chapter we focused exclusively on the methods to predict domain-domain and domain-peptide interactions that use, in a significant way, high-throughput protein interaction networks. Within this group of methods we selected a set (by no means exhaustive) of representative approaches. We demonstrated that, despite the fact that high-throughput interactions are inherently noisy, they provide extremely valuable resource for predicting domain and peptide interactions. The noise in the high-throughput protein interaction data dictates, however, that the methods that are based exclusively on the network information are only capable of predicting interactions occurring multiple times.

An important and not completely resolved problem is the issue of evaluation of prediction methods. A standard way to assess such methods is to test how well they predict known interactions. Yet, the set of currently known interactions is not only very limited but since PDB data is well known to be biased (Brenner et al. 1997; Gerstein 1998; Peng et al. 2004; Mestres 2005; Xie and Bourne 2005) such biases are also likely to be inherited by iPFAM (Guimarães and Przytycka 2008). For example, in the context of domain-domain interactions, the crystal structures favor stable and well studied protein complexes. Therefore, an important issue in prediction methods is an experimental validation of new predicted interactions.

References

Apic, G., J. Gough, et al. (2001). "An insight into domain combinations." *Bioinformatics* **17 Suppl 1**: S83–9.

Bock, J. R. and D. A. Gough (2001). "Predicting protein–protein interactions from primary structure." *Bioinformatics* **17**(5): 455–60.

Brenner, S. E., C. Chothia, et al. (1997). "Population statistics of protein structures: lessons from structural classifications." *Curr Opin Struct Biol* **7**(3): 369–76.

Cormen, T. H., C. H. Leiserson, et al. (2001). *Introduction to Algorithms*, MIT Press.

Davey, N. E., D. C. Shields, et al. (2006). "SLiMDisc: short, linear motif discovery, correcting for common evolutionary descent." *Nucleic Acids Res* **34**(12): 3546–54.

Deng, M., S. Mehta, et al. (2002). "Inferring domain-domain interactions from protein-protein interactions." *Genome Res* **12**(10): 1540–8.

Ferraro, E., A. Via, et al. (2006). "A novel structure-based encoding for machine-learning applied to the inference of SH3 domain specificity." *Bioinformatics* **22**(19): 2333–9.

Finn, R. D., J. Mistry, et al. (2006). "Pfam: clans, web tools and services." *Nucleic Acids Res* **34**(Database issue): D247–51.

Gerstein, M. (1998). "Patterns of protein-fold usage in eight microbial genomes: a comprehensive structural census." *Proteins* **33**(4): 518–34.

Gomez, S. M. and A. Rzhetsky (2002). "Towards the prediction of complete protein–protein interaction networks." *Pac Symp Biocomput*: 413–24.

Guimarães, K. S., R. Jothi, et al. (2006). "Predicting domain-domain interactions using a parsimony approach." *Genome Biol* **7**(11): R104.

Guimarães, K. S. and T. M. Przytycka (2008). "Interrogating domain-domain interactions with parsimony based approaches." *BMC Bioinformatics* **9**:171

Harris, M. A., J. Clark, et al. (2004). "The Gene Ontology (GO) database and informatics resource." *Nucleic Acids Res* **32**(Database issue): D258–61.

Hjerrild, M., A. Stensballe, et al. (2004). "Identification of Phosphorylation Sites in Protein Kinase A Substrates Using Artificial Neural Networks and Mass Spectrometry." *J Proteome Res* **3**(3): 426–433.

Hulo, N., A. Bairoch, et al. (2006). "The PROSITE database." *Nucleic Acids Res* **34**(Database issue): D227–30.

Itzhaki, Z., E. Akiva, et al. (2006). "Evolutionary conservation of domain-domain interactions." *Genome Biol* **7**(12): R125.

Jothi, R., P. F. Cherukuri, et al. (2006). "Co-evolutionary analysis of domains in interacting proteins reveals insights into domain-domain interactions mediating protein-protein interactions." *J Mol Biol* **362**(4): 861–75.

Kanehisa, M., S. Goto, et al. (2006). "From genomics to chemical genomics: new developments in KEGG." *Nucleic Acids Res* **34**(Database issue): D354–57.

Kann, M. G., R. Jothi, et al. (2007). "Predicting protein domain interactions from coevolution of conserved regions." *Proteins* **67**(4): 811–20.

Krebs, E. G. and J. A. Beavo (1979). "Phosphorylation-dephosphorylation of enzymes." *Annu Rev Biochem* **48**: 923–59.

Lander, E. S., L. M. Linton, et al. (2001). "Initial sequencing and analysis of the human genome." *Nature* **409**(6822): 860–921.

Lee, H., M. Deng, et al. (2006). "An integrated approach to the prediction of domain-domain interactions." *BMC Bioinformatics* **7**: 269.

Lehrach, W. P., D. Husmeier, et al. (2006). "A regularized discriminative model for the prediction of protein-peptide interactions." *Bioinformatics* **22**(5): 532–40.

Linding, R., L. J. Jensen, et al. (2007). "Systematic discovery of in vivo phosphorylation networks." *Cell* **129**(7): 1415–26.

Marcotte, E. M., M. Pellegrini, et al. (1999). "Detecting protein function and protein-protein interactions from genome sequences." *Science* **285**(5428): 751–53.

Mestres, J. (2005). "Representativity of target families in the Protein Data Bank: impact for family-directed structure-based drug discovery." *Drug Discov Today* **10**(23–24): 1629–37.

Neduva, V., R. Linding, et al. (2005). "Systematic discovery of new recognition peptides mediating protein interaction networks." *PLoS Biol* **3**(12): e405.

Ng, S. K., Z. Zhang, et al. (2003a). "Integrative approach for computationally inferring protein domain interactions." *Bioinformatics* **19**(8): 923–9.

Ng, S. K., Z. Zhang, et al. (2003b). "InterDom: a database of putative interacting protein domains for validating predicted protein interactions and complexes." *Nucleic Acids Res* **31**(1): 251–4.

Nguyen, T. P. and T. B. Ho (2006). "Discovering signal transduction networks using signaling domain-domain interactions." *Genome Inform* **17**(2): 35–45.

Nye, T. M., C. Berzuini, et al. (2005). "Statistical analysis of domains in interacting protein pairs." *Bioinformatics* **21**(7): 993–1001.

Obenauer, J. C., L. C. Cantley, et al. (2003). "Scansite 2.0: proteome-wide prediction of cell signaling interactions using short sequence motifs." *Nucleic Acids Res* **31**(13): 3635–41.

Pagel, P., P. Wong, et al. (2004). "A domain interaction map based on phylogenetic profiling." *J Mol Biol* **344**(5): 1331–46.

Pawson, T. and P. Nash (2003). "Assembly of cell regulatory systems through protein interaction domains." *Science* **300**(5618): 445–52.

Peng K., Z. Obradovic, et al. (2004). "Exploring bias in the Protein Data Bank using contrast classifiers." *Pac Symp Biocomput*: 435–446.

Peri, S., J. D. Navarro, et al. (2003). "Development of human protein reference database as an initial platform for approaching systems biology in humans." *Genome Res* **13**(10): 2363–71.

Ptacek, J., G. Devgan, et al. (2005). "Global analysis of protein phosphorylation in yeast." *Nature* **438**(7068): 679–84.

Puntervoll, P., R. Linding, et al. (2003). "ELM server: a new resource for investigating short functional sites in modular eukaryotic proteins." *Nucl Acids Res.* **31**(13): 3625–3630.

Reiss, D. J. and B. Schwikowski (2004). "Predicting protein-peptide interactions via a network-based motif sampler." *Bioinformatics* **20 Suppl 1**: i274–82.

Rigoutsos, I. and A. Floratos (1998). "Combinatorial pattern discovery in biological sequences: the TEIRESIAS algorithm." *Bioinformatics* **14**(1): 55–67.

Riley, R., C. Lee, et al. (2005). "Inferring protein domain interactions from databases of interacting proteins." *Genome Biol* **6**(10): R89.

Shmulevich, I., E. R. Dougherty, et al. (2002). "Probabilistic Boolean Networks: a rule-based uncertainty model for gene regulatory networks." *Bioinformatics* **18**(2): 261–74.

Shoemaker, B. A., A. R. Panchenko, et al. (2006). "Finding biologically relevant protein domain interactions: conserved binding mode analysis." *Protein Sci* **15**(2): 352–61.

Singhal, M. and H. Resat (2007). "A domain-based approach to predict protein-protein interactions." *BMC Bioinformatics* **8**: 199.

Sprinzak, E. and H. Margalit (2001). "Correlated sequence-signatures as markers of protein-protein interaction." *J Mol Biol* **311**(4): 681–92.

Stein, A., R. B. Russell, et al. (2005). "3did: interacting protein domains of known three-dimensional structure." *Nucleic Acids Res* **33**(Database issue): D413–7.

Suthram, S., T. Shlomi, et al. (2006). "A direct comparison of protein interaction confidence assignment schemes." *BMC Bioinformatics* **7**: 360.

Toro, I., S. Thore, et al. (2001). "RNA binding in an Sm core domain: X-ray structure and functional analysis of an archaeal Sm protein complex." *Embo J* **20**(9): 2293–303.

von Mering, C., L. J. Jensen, et al. (2007). "STRING 7 – Recent developments in the integration and prediction of protein interactions." *Nucleic Acids Research* **35**: D358–D362.

Wang, H., E. Segal, et al. (2007). "InSite: a computational method for identifying protein-protein interaction binding sites on a proteome-wide scale." *Genome Biol* **8**(9): R192.

Wojcik, J. and V. Schachter (2001). "Protein-protein interaction map inference using interacting domain profile pairs." *Bioinformatics* **17 Suppl 1**: S296–305.
Xie, L. and P. E. Bourne (2005). "Functional coverage of the human genome by existing structures, structural genomics targets, and homology models." *PLoS Comput Biol* **1**(3): e31.
Zhang, L. V., S. L. Wong, et al. (2004). "Predicting co-complexed protein pairs using genomic and proteomic data integration." *BMC Bioinformatics* **5**: 38.

Chapter 6
Integrative Structure Determination of Protein Assemblies by Satisfaction of Spatial Restraints

Frank Alber, Brian T. Chait, Michael P. Rout, and Andrej Sali

Abstract To understand the cell, we need to determine the structures of macromolecular assemblies, many of which consist of tens to hundreds of components. A great variety of experimental data can be used to characterize the assemblies at several levels of resolution, from atomic structures to component configurations. To maximize completeness, resolution, accuracy, precision and efficiency of the structure determination, a computational approach is needed that can use spatial information from a variety of experimental methods. We propose such an approach, defined by its three main components: a hierarchical representation of the assembly, a scoring function consisting of spatial restraints derived from experimental data, and an optimization method that generates structures consistent with the data. We illustrate the approach by determining the configuration of the 456 proteins in the nuclear pore complex from Baker's yeast.

6.1 Introduction

Assemblies as functional modules of the cell. Macromolecular assemblies consist of non-covalently interacting macromolecular components, such as proteins and nucleic acids. They vary widely in size and play crucial roles in most cellular processes (Alberts 1998). Many assemblies are composed of tens and even hundreds of individual components. For example, the nuclear pore complex (NPC) of ~456 proteins regulates macromolecular transport across the nuclear envelope (NE); the ribosome consists of ~80 proteins and ~15 RNA molecules and is responsible for protein biosynthesis.

Need for assembly structures. A comprehensive characterization of the structures and dynamics of biological assemblies is essential for a mechanistic understanding of the cell (Alber et al. 2008; Robinson et al. 2007; Sali 2003; Sali et al. 2003; Sali and Kuriyan 1999). Even a coarse characterization of the configuration of macromolecular components in a complex (Fig. 6.1) helps to elucidate the principles that

F. Alber
Molecular and Computational Biology Program, Department of Biological Sciences, University of Southern California, Los Angeles, CA 90089-1340, USA
e-mail: alber@usc.edu

A. Panchenko, T. Przytycka (eds.), *Protein-protein Interactions and Networks*,
DOI: 10.1007/978-1-84800-125-1_6,

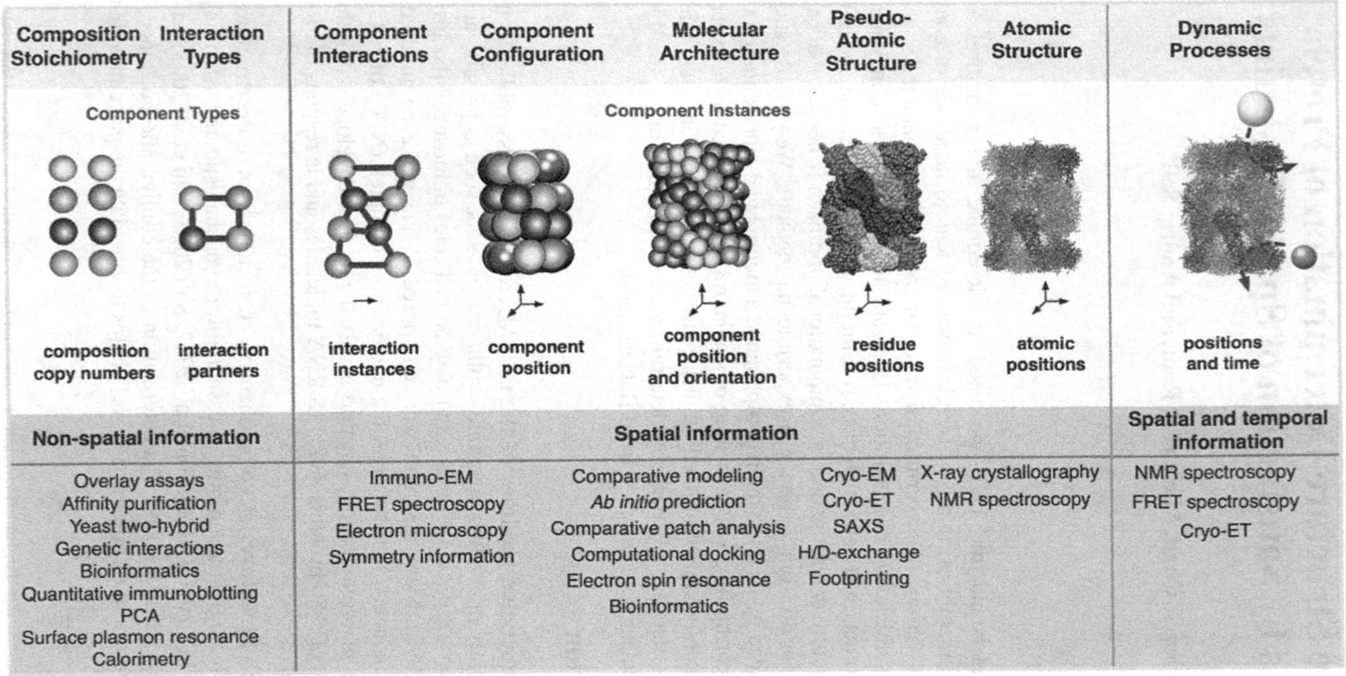

Fig. 6.1 Structural information about an assembly (Alber et al. 2008). Varied experimental methods can determine the copy numbers (stoichiometry) and types (composition) of the components (whether or not components interact with each other), positions of the components, and their relative orientations. Importantly, some methods identify only component types and do not distinguish between different instances of a component of the same type when more than one copy of it is present in the assembly. Other methods do identify specific instances of a component. Integration of data from varied methods generally increases the accuracy, efficiency, and coverage of structure determination. PCA, protein-fragment complementation assay; H/D, hydrogen/deuterium; EM, electron microscopy; FRET, fluorescence resonance energy transfer; SAXS, small angle X-ray scattering

underlie cellular processes, in addition to providing a necessary starting point for a higher resolution description.

Scope. Complete lists of macromolecular components of biological systems are becoming available (Aebersold and Mann 2003). However, the identification of complexes between these components is a non-trivial task. This difficulty arises partly from the multitude of component types and the varying lifespan of the complexes (Russell et al. 2004). The most comprehensive information about binary protein interactions is available for the *Saccharomyces cerevisiae* proteome, consisting of ~6,200 proteins. This data has been generated by methods such as the yeast two-hybrid system (Ito et al. 2000; Uetz et al. 2000) and affinity purifications coupled with mass-spectrometry (Collins et al. 2007; Gavin et al. 2006; Krogan et al. 2006). The lower bound on binary protein interactions in yeast has been estimated to be ~30,000 (Russell et al. 2004), corresponding to the average of ~9 protein partners per protein, though not necessarily all at the same time. The number of higher order complexes in yeast is estimated to be ~800, based on affinity purification experiments (Collins et al. 2007; Devos and Russell 2007; Gavin et al. 2006; Krogan et al. 2006). The human proteome may have an order of magnitude more complexes than the yeast cell; and the number of different complexes across all relevant genomes may be several times larger still. Therefore, there may be thousands of biologically relevant macromolecular complexes in a few hundred key cellular processes whose stable structures and transient interactions are yet to be characterized (Abbott 2002; Alberts 1998).

Difficulties. Compared to structure determination of the individual components, however, structural characterization of macromolecular assemblies is usually more difficult and represents a major challenge in structural biology (Alber et al. 2008; Robinson et al. 2007; Sali et al. 2003; Sali and Kuriyan 1999). For example, X-ray crystallography is limited by the ability to grow suitable crystals and to build molecular models into large unit cells; nuclear magnetic resonance (NMR) spectroscopy is limited by size; electron microscopy (EM), affinity purification, yeast two-hybrid system, calorimetry, footprinting, chemical cross-linking, small angle X-ray scattering (SAXS), and fluorescence resonance energy transfer (FRET) spectroscopy are limited by low resolution of the corresponding structural information; and computational protein structure modeling and docking are limited by low accuracy.

Integrative approach. These shortcomings can be minimized by simultaneous consideration of all available information about a given assembly (Fig. 6.1) (Alber et al. 2007a; Alber et al. 2004; Alber et al. 2008; Harris et al. 1994; Malhotra and Harvey 1994; Robinson et al. 2007; Sali et al. 2003). This information may vary greatly in terms of its accuracy and precision, and includes data from both experimental methods and theoretical considerations, such as those listed above. The integration of structural information about an assembly from various sources can only be achieved by computational means. In this review, we focus on the computational aspects of this data integration.

Review outline. We begin by listing the types of spatial information generated by experimental and computational methods that have allowed structural biology to shift its focus from individual proteins to large assemblies. Next, we offer a perspective on generating macromolecular assemblies that are consistent with all available information from experimental methods, physical theories, and statistical preferences extracted from biological databases. Such an integrative system in principle achieves higher completeness, resolution, accuracy, precision, and efficiency than a structure characterization based on any of the individual types of data alone (Alber et al. 2007a; Alber et al. 2008; Robinson et al. 2007; Sali et al. 2003). Finally, we illustrate this approach by its application to the determination of the configuration of 456 proteins in the yeast NPC (Alber et al. 2007a; Alber et al. 2007b).

6.2 Sources of Spatial Information

Different experimental methods produce different types of structural information (Fig. 6.1). This information varies in terms of what spatial features it restrains as well as in resolution, accuracy, and quantity. The stoichiometry and composition of protein components in an assembly can be determined by methods such as quantitative immunoblotting and mass spectrometry. The positions of the components can be elucidated by cryo-EM and labeling techniques. Whether or not components interact with each other can be measured by yeast two-hybrid system and affinity purification. Relative orientations of components and information about interacting residues can be inferred from cryo-EM, hydrogen/deuterium (H/D) exchange, OH radical footprinting, and chemical-crosslinking. At the highest resolution, information about the atomic structures of components and their interactions can be determined by X-ray crystallography and NMR spectroscopy.

Importantly, some methods do not distinguish between different instances of a component of the same type, resulting in ambiguity when more than one copy of the component is present in the assembly (e.g., proteomics methods, including yeast two hybrid system and affinity purification). Structures can be described at different levels of resolution, including the component configuration (specifying component positions and the presence of interactions), the molecular architecture (specifying the components' configuration and relative orientations), pseudo-atomic models (specifying atomic positions with errors larger than the size of an atom), and atomic structures (specifying atomic positions with precision smaller than the size of an atom).

6.3 Comprehensive Data Integration by Satisfaction of Spatial Restraints

The experimental data about a structure must be converted to an explicit structural model through computation. Importantly, these computational methods differ in the type of information they use to calculate the assembly structures, rather than how they calculate them once the information is specified.

Detailed structural characterization of assemblies is often difficult by any single existing experimental or computational method. We suggest that this barrier can be overcome by "hybrid" approaches that integrate data from diverse biochemical and biophysical experiments as well as computational methods. This information may vary greatly in terms of its resolution, accuracy, and quantity. Here, we outline an approach for generating structures of macromolecular assemblies that are consistent with all available information from experimental methods, physical theories, and statistical preferences extracted from biological databases. Such an integrative system will help to maximize efficiency, resolution, accuracy, precision, and completeness of the structural coverage of macromolecular assemblies.

In this section, we describe the underlying theory and methods of our hybrid approach to characterizing macromolecular assembly structures. A sample application is provided by the structure determination of the NPC (Alber et al. 2007a; Alber et al. 2007b; Alber et al. 2004; Alber et al. 2005; Sali and Kuriyan 1999) (below).

Formalization of the problem. The complete process of structure determination can be seen as a potentially iterative series of four steps, including data generation by experiments, data translation into spatial restraints, calculation of an ensemble of structures by satisfaction of spatial restraints, and an analysis of the ensemble. The structural characterization part of the process can be expressed as an optimization problem (Fig. 6.2). In this view, models that are consistent with the input

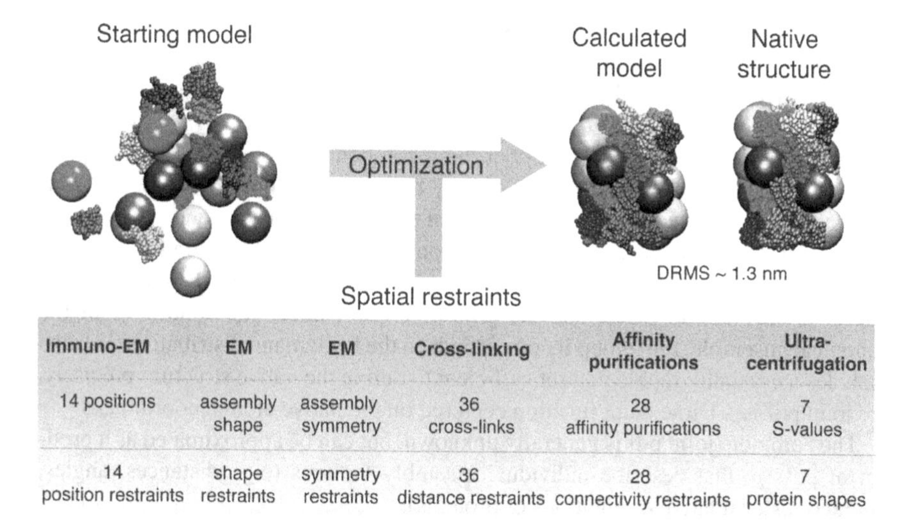

Fig. 6.2 Characterization of an assembly configuration based on data simulated from a known native structure (Alber et al. 2008). In this approach (Alber et al. 2005), the simulated data include protein positions (e.g., from immuno-EM), assembly shape (e.g., from EM), relative proximity of components (e.g., from cross-linking and affinity purification). The data is translated into spatial restraints that are then summed to obtain a scoring function. A random starting structure is optimized by a combination of conjugate gradients and molecular dynamics with simulated annealing to minimize violations of all restraints. The listed data was sufficient to identify the coarse relative position of each protein (i.e., the protein configuration). To illustrate the possibility of using different representations for different proteins, a protein is represented either by an X-ray structure or by a single sphere that best reproduces its hydrodynamic properties determined by ultracentrifugation. DRMS, distance root-mean-square difference between the protein centroids in the determined model and the native structure

information are calculated by optimizing a scoring function. The three components of this approach are (i) a representation of the modeled assembly, (ii) a scoring function consisting of the individual spatial restraints, and (iii) optimization of the scoring function to obtain all possible models that satisfy the input restraints.

Representation. The modeled structure is represented by a hierarchy of particles, defined by their positions and other properties (Fig. 6.2). For a protein assembly, the hierarchy can include atoms, atomic groups, amino acid residues, secondary structure segments, domains, proteins, protein sub-complexes, symmetry units, and the whole assembly. The coordinates and properties of particles at any level are calculated from those at the highest resolution level. Different parts of the assembly can be represented at different resolutions to reflect the input information about the structure (Fig. 6.2). Moreover, different representations can also apply to the same part of the system. For example, affinity purification may indicate proximity between two proteins and cross-linking may indicate which specific residues are involved in the interaction.

Scoring Function. The most important aspect of structure characterization is to accurately capture all experimental, physical, and statistical information about the modeled structure. This objective is achieved by expressing our knowledge of any kind as a scoring function whose global optimum corresponds to the native assembly structure (Shen and Sali 2006). One such function is a joint probability density function (pdf) of the Cartesian coordinates of all assembly proteins, given the available information I about the system, $p(C|I)$. $C = (c_1, c_2, \ldots, c_n)$ is the list of the Cartesian coordinates (c_1) of the n component proteins in the assembly. The joint pdf p gives the probability density that a component i of the native configuration is positioned very close to c_i, given the information I we wish to consider in the calculation. In general, I may include any structural information from experiments, physical theories, and statistical preferences. For example, when information I reflects only the sequence and the laws of physics under the conditions of the canonical ensemble, the joint pdf corresponds to the Boltzmann distribution. If I also includes a crystallographic dataset sufficient to define the native structure precisely, the joint pdf is a Dirac delta function centered on the native atomic coordinates.

The complete joint pdf is generally unknown, but can be approximated as a product of pdfs p_f that describe individual assembly features (e.g., distances, angles, interactions or relative orientations of proteins):

$$p(C\,|I) = \prod_f p_f(C|I_f)$$

The scoring function *F(C)* is then defined as the logarithm of the joint pdf:

$$F(C) = -\ln \prod_f p_f(C|I_f) = \sum_f r_f(C)$$

For convenience, we refer to the logarithm of a feature pdf as a restraint r_f and the scoring function is therefore a sum of the individual restraints.

Restraints. A restraint r_f can in principle have any functional form. However, it is convenient if ideal solutions consistent with the data correspond to values of 0, while values larger than 0 correspond to a violated restraint; for example, a restraint is frequently a harmonic function of the restrained feature.

Restrained features. The restrained features in principle include any structural aspect of an assembly, such as contacts, proximity, distances, angles, chirality, surface, volume, excluded volume, shape, symmetry, and localization of particles and sets of particles.

Translating data into restraints. A key challenge is to accurately express the input data and their uncertainties in terms of the individual spatial restraints. An interpretation of the data in terms of a spatial restraint generally involves identifying the restrained components (i.e., structural interpretation) and the possible values of the restrained feature implied by the data. For instance, the shape, density and symmetry of a complex or its subunits may be derived from X-ray crystallography and EM (Frank 2006); upper distance bounds on residues from different proteins may be obtained from NMR spectroscopy (Fiaux et al. 2002) and chemical cross-linking (Trester-Zedlitz et al. 2003); protein-protein interactions may be discovered by the yeast two-hybrid system (Phizicky et al. 2003) and calorimetry (Lakey and Raggett 1998); two proteins can be assigned to be in proximity if they are part of an isolated sub-complex identified by affinity purification in combination with mass spectrometry (Bauer and Kuster 2003). Increasingly, important restraints will be derived from pairwise molecular docking (Mendez et al. 2005), statistical preferences observed in the structurally defined protein-protein interactions (Davis and Sali 2005), and analysis of multiple sequence alignments (Valencia and Pazos 2002).

Conditional restraints. If structural interpretation of the data is ambiguous (i.e., the data cannot be uniquely assigned to specific components), only "conditional restraints" can be defined. For example, when there is more than one copy of a protein per assembly, a yeast two-hybrid system indicates only which protein types but not which instances interact with each other. Such ambiguous information must be translated into a conditional restraint that considers all alternative structural interpretations of the data (Fig. 6.3). The selection of the best alternative interpretation is then achieved as part of the structure optimization process.

Figure 6.3 shows a conditional restraint that encodes protein contacts consistent with an affinity purification experiment (Alber et al. 2007a; Alber et al. 2008; Alber et al. 2005) (Fig. 6.3). In this example, affinity purification identified 4 protein types (yellow, blue, red, green), derived from an assembly containing a single copy of the yellow, blue, and red protein and two copies of the green protein. The sample affinity purification implies that at least 3 of the following 6 possible types of interaction must occur: blue-red, blue-yellow, blue-green, red-green, red-yellow, and yellow-green. In addition, (i) the three selected interactions must form a spanning tree of the composite graph (Fig. 6.3); (ii) each type of interaction can involve either copy of the green protein; and (iii) each protein can interact through any of its beads. These considerations can be encoded through a tree-like evaluation of the conditional restraint. At the top level, all possible bead-bead interactions between all protein copies are clustered by protein types. Each alternative bead interaction

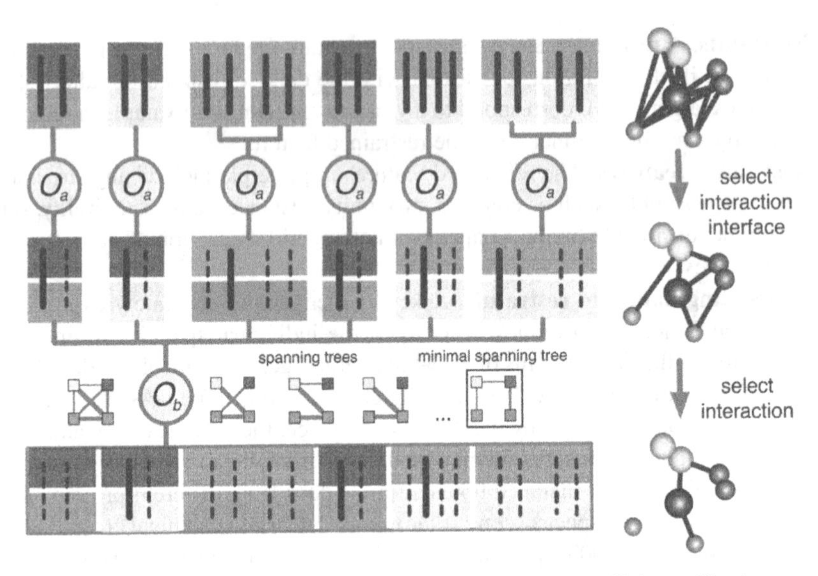

Fig. 6.3 Conditional restraint encoding protein contacts based on an affinity purification experiment that identified 4 protein types (*yellow*, *blue*, *red*, *green*), derived from an assembly containing a single copy of the *yellow*, *blue*, and *red* protein and two copies of the *green* protein (Alber et al. 2007a; Alber et al. 2008; Alber et al. 2005). A single protein is represented by either one bead (*blue* and *green* proteins) or two beads (*yellow* and *red* proteins) (column on the *right*); alternative interactions between proteins are indicated by different edges. Protein contacts are selected in a decision tree-like evaluation process by operator functions O_a and O_b (*left panel*) (see text for a detailed description). *Red vertical lines* indicate restraints that encode a protein contact; *thick vertical lines* are a subset of restraints that are selected for contribution to the final value of the conditional restraint, whereas *dotted vertical lines* indicate restraints that are not selected. Also shown are spanning trees of a "composite graph". The composite graph is a fully connected graph that consists of nodes for all identified protein types (square nodes) and edges for all pairwise interactions between protein types (*left* to the O_b operator); edge weights correspond to the violations of interaction restraints and quantify how consistent is the corresponding interaction with the current assembly structure. A "spanning tree" is a graph with the smallest possible number of edges that connect all nodes; a subset of 4 out of 16 spanning trees is indicated to the *right* of the O_b operator. The "minimal spanning tree" is the spanning tree with the minimal sum of edge weights (i.e., restraints violations)

can be restrained by a restraint corresponding to a harmonic upper bound on the distance between the beads; these are termed "optional restraints", because only a subset is selected for contribution to the final value of the conditional restraint. Next, an operator function (O_a) selects only the least violated optional restraint from each interaction type, resulting in 5 restraints (thick red line) at the middle level of the tree (Fig. 6.3). Finally, a minimal spanning tree operator (O_b) finds the minimal spanning tree corresponding to the combination of 3 restraints that are most consistent with the affinity purification (thick red lines in Fig. 6.3). The whole restraint evaluation process is executed at each optimization step based on the current configuration, thus resulting in possibly different subsets of selected optional restraints at each optimization step.

Optimization methods. Structures can be generated by simultaneously minimizing the violations of all restraints, resulting in configurations that minimize the scoring function *F*. It is crucial to have access to multiple optimization methods to choose one that works best with a specific scoring function and representation. Optimization methods implemented in our program IMP currently include conjugate gradients, quasi-Newton minimization, and molecular dynamics, as well as more sophisticated schemes, such as self-guided Langevin dynamics, the replica exchange method, and exact inference (belief propagation) (K. Lasker, M. Topf, A. Sali and H. Wolfson, unpublished information); all of these methods can refine positions of the individual particles as well as treat subsets of particles as rigid bodies.

Outcomes. There are three possible outcomes of the calculation. First, if only a single model satisfies all input information, there is probably sufficient data for prediction of the unique native state. Second, if different models are consistent with the input information, the data are insufficient to define the single native state or there are multiple native structures. If the number of distinct models is small, the structural differences between the models may suggest additional experiments to narrow down the possible solutions. Third, if no models satisfy all input information, the data or their interpretation in terms of the restraints are incorrect.

Analysis. In general, a number of different configurations may be consistent with the input restraints. The aim is to obtain as many structures as possible that satisfy all input restraints. To comprehensively sample such structural solutions consistent with the data, independent optimizations of randomly generated initial configurations need to be performed until an "ensemble" of structures satisfying the input restraints is obtained. The ensemble can then be analyzed in terms of assembly features, such as the protein positions, contacts, and configuration. These features can generally vary among the individual models in the ensemble. To analyze this variability, a probability distribution of each feature can be calculated from the ensemble. Of particular interest are the features that are present in most configurations in the ensemble and have a single maximum in their probability distribution. The spread around the maximum describes how precisely the feature was determined by the input restraints. When multiple maxima are present in the feature distribution at the precision of interest, the input restraints are insufficient to define the single native state of the corresponding feature (or there are multiple native states).

Predicting Accuracy. Assessing the accuracy of a structure is important and difficult. The accuracy of a model is defined as the difference between the model and the real native structure. Therefore, it is impossible to know with certainty the accuracy of the proposed structure, without knowing the real native structure. Nevertheless, our confidence can be modulated by five considerations: (i) self-consistency of independent experimental data; (ii) structural similarity among all configurations in the ensemble that satisfy the input restraints; (iii) simulations where a native structure is assumed, corresponding restraints simulated from it, and the resulting calculated structure compared with the assumed native structure; (iv) confirmatory spatial data that were not used in the calculation of the structure (e.g., criterion similar to the crystallographic free R-factor (Brunger 1993) can be used to assess both the model accuracy and the harmony among the input restraints); and (v) patterns emerging from a mapping of independent and unused data on the structure that are unlikely to occur by chance (Alber et al. 2007a; Alber et al. 2007b).

Advantages. The integrative approach to structure determination has several advantages: (i) It benefits from the synergy among the input data, minimizing the drawback of incomplete, inaccurate, and/or imprecise data sets (although each individual restraint may contain little structural information, the concurrent satisfaction of all restraints derived from independent experiments may drastically reduce the degeneracy of structural solutions); (ii) it can potentially produce all structures that are consistent with the data, not just one; (iii) the variation among the structures consistent with the data allows us to assess sufficiency of the data and the precision of the representative structure; (iv) it can make the process of structure determination more efficient by indicating what measurements would be most informative.

6.4 Structural Characterization of the Nuclear Pore Complex

Using the approach outlined above, we determined the native configuration of proteins in the yeast nuclear pore complex (NPC) (Alber et al. 2007a; Alber et al. 2007b). NPCs are large (~50 MDa) proteinaceous assemblies spanning the nuclear envelope (NE), where they function as the sole mediators of bidirectional exchange between the nucleoplasmic and cytoplasmic compartments in all eukaryotes (Lim and Fahrenkrog 2006). EM images of the yeast NPC at ~200 Å resolution revealed that the nuclear pore forms a channel by stacking two similar rings, each one consisting of 8 radially arranged “half-spoke” units (Yang et al. 1998). The yeast NPC is build from multiple copies of 30 different proteins, totaling approximately 456 proteins (nups).

Although low-resolution EM has provided valuable insights into the overall shape of the NPC, the spatial configuration of its component proteins and the detailed interaction network between them was unknown. A description of the NPC’s structure was needed to understand its function and assembly, and to provide clues to its evolutionary origins. Due to its size and flexibility, detailed structural characterization of the complete NPC assembly has proven to be extraordinarily challenging. Further compounding the problem, atomic structures have only been solved for domains covering ~5% of the protein sequence (Devos et al. 2006).

To determine the protein configuration of the NPC, we collected a large and diverse set of biophysical and biochemical data. The data was derived from six experimental sources (Fig. 6.4): (i) Quantitative immuno-blotting experiments determined the stoichiometry of all 30 nups in the NPC; (ii) hydrodynamics experiments provided information about the approximate excluded volume and the coarse shape of each nup; (iii) immuno-EM provided a coarse localization for each nup along two principal axes of the NPC; (iv) an exhaustive set of affinity purification experiments determined the composition of 77 NPC complexes; (v) overlay experiments determined 5 direct binary nup interactions; and (vi) symmetry considerations and the dimensions of the NE were extracted from cryo-EM. Moreover,

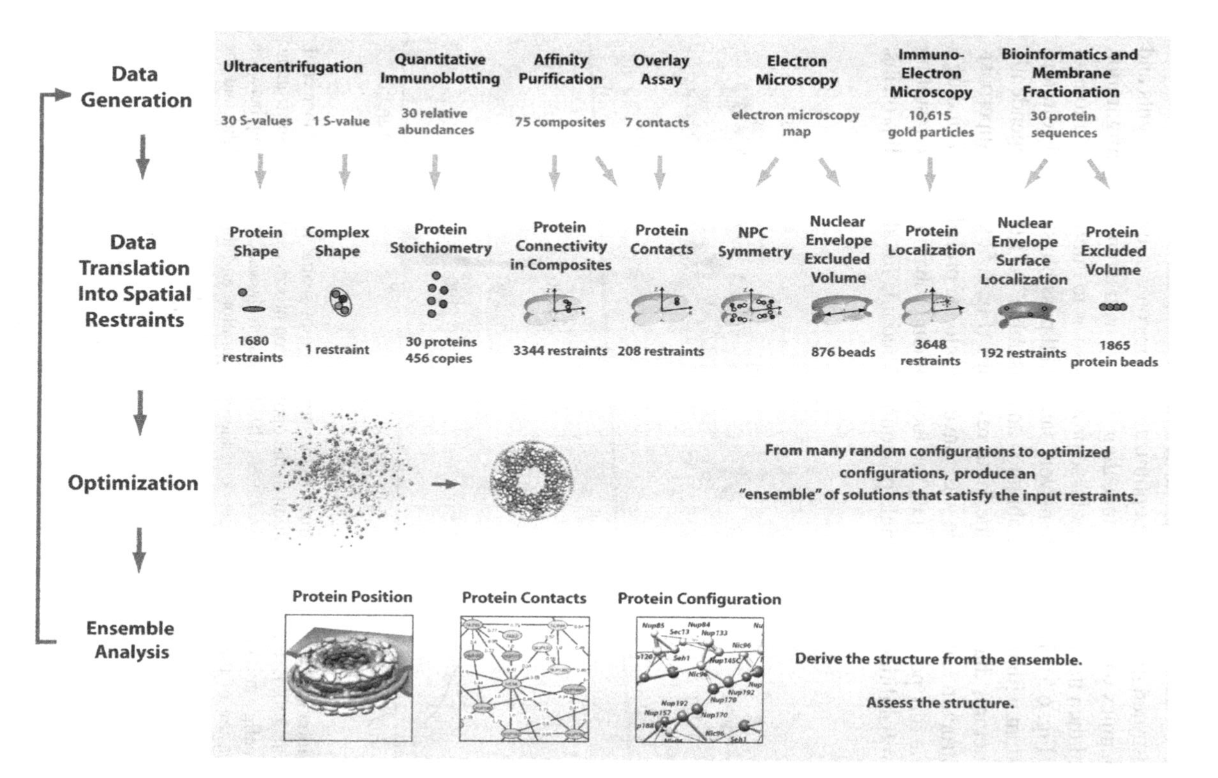

Fig. 6.4 Determining the architecture of the NPC by integrating spatial restraints from proteomic data (Alber et al. 2007a). First, structural data (*red*) are generated by various experiments (*black*). Second, the data and theoretical considerations are expressed as spatial restraints (*blue*). Third, an ensemble of structural solutions that satisfy the data is obtained by minimizing the violations of the spatial restraints, starting from many different random configurations. Fourth, the ensemble is clustered into sets of distinct solutions as well as analyzed in terms of protein positions, contacts, and configuration

bioinformatics analysis provided information about the position of transmembrane helices for the three integral membrane nups. This data was translated into spatial restraints on the NPC (Fig. 6.4).

The relative positions and proximities of the NPC's constituent proteins were then produced by satisfying these spatial restraints, using the approach described above and illustrated in Fig. 6.5. The optimization relies on conjugate gradients and molecular dynamics with simulated annealing. It starts with a random configuration of proteins and then iteratively moves these proteins so as to minimize violations of the restraints (Fig. 6.5). To comprehensively sample all possible structural solutions that are consistent with the data, we obtained an "ensemble" of 1,000 independently calculated structures that satisfied the input restraints (Fig. 6.5c). After superposition of these structures, the ensemble was converted into the probability of finding a given protein at any point in space (i.e., the localization probability). The resulting localization probabilities yielded single pronounced maxima for almost all proteins,

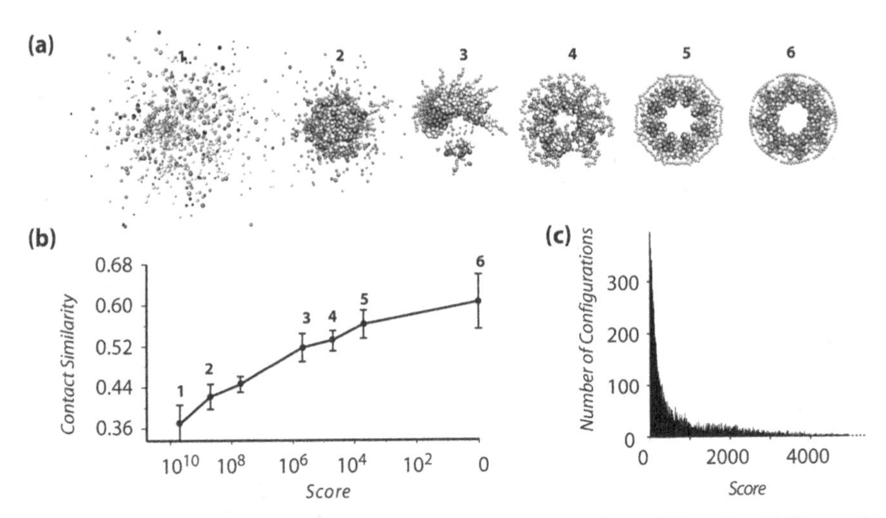

Fig 6.5 Calculation of the NPC bead structure by satisfaction of spatial restraints (Alber et al. 2007a; Alber et al. 2008). (**a**), Representation of the optimization process as it progresses from an initial random configuration to an optimal solution. (**b**), The graph shows the relationship between the score (a measure of the consistency between the configuration and the input data) and the average contact similarity. The contact similarity quantifies how similar two configurations are in terms of the number and types of their protein contacts; two proteins are considered to be in contact when they are sufficiently close to one another given their size and shape. The average contact similarity at a given score is determined from the contact similarities between the lowest scoring configuration and a sample of 100 configurations with the given score. Error bars indicate standard deviation. Representative configurations at various stages of the optimization process from *left* (very large scores) to *right* (with a score of 0) are shown above the graph; a score of 0 indicates that all input restraints have been satisfied. As the score approaches zero, the contact similarity increases, showing that there is only a single cluster of closely related configurations that satisfy the input data. (**c**), Distribution of configuration scores demonstrates that our sampling procedure finds configurations consistent with the input data. These configurations satisfy all the input restraints within the experimental error

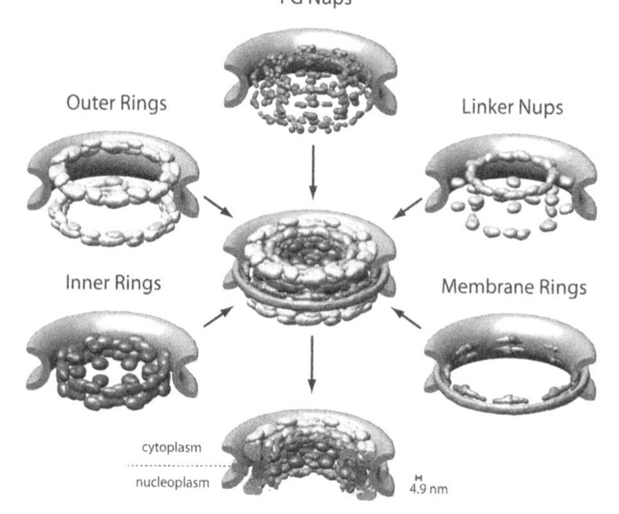

Fig. 6.6 Localization of major substructures and their component proteins in the NPC (Alber et al. 2007b; Alber et al. 2008). The proteins are represented by their localization volumes (Alber et al. 2007a) and have been colored according to their classification into five distinct substructures based on their location and functional properties: the outer rings in *yellow*, the inner rings in *purple*, the membrane rings in *brown*, the linker nups in *blue* and *pink*, and the FG nups (for which only the structured domains are shown) in *green*. The pore membrane is shown in *gray*

demonstrating that the input restraints define a single NPC architecture (Fig. 6.6). The average standard deviation for the separation between neighbouring protein centroids is 5 nm. Given that this level of precision is less than the diameter of many proteins, our map is sufficient to determine the relative position of proteins in the NPC. Although each individual restraint may contain little structural information, the concurrent satisfaction of all restraints derived from independent experiments drastically reduces the degeneracy of the structural solutions (Fig. 6.7).

Our structure (Fig. 6.5) reveals that half of the NPC is made of a core scaffold, which is structurally analogous to vesicle coating complexes. This scaffold forms an interlaced network that coats the entire curved surface of the NE within which the NPC is embedded. The selective barrier for transport is formed by large numbers of proteins with disordered regions that line the inner face of the scaffold. The NPC consists of only a few structural modules. These modules resemble each other in terms of the configuration of their homologous constituents. The architecture of the NPC thus appears to be based on the hierarchical repetition of the modules that likely evolved through a series of gene duplications and divergences. Thus, the determination of the NPC configuration in combination with the fold prediction of its constituent proteins (Devos 2004, 2006) can provide clues to the ancient evolutionary origins of the NPC.

In the future, we envision combining electron tomography, proteomics, cross-linking, cryo-EM of subcomplexes, and experimentally determined or modeled atomic structures of the individual subunits to obtain a pseudo-atomic model of the whole NPC assembly in action.

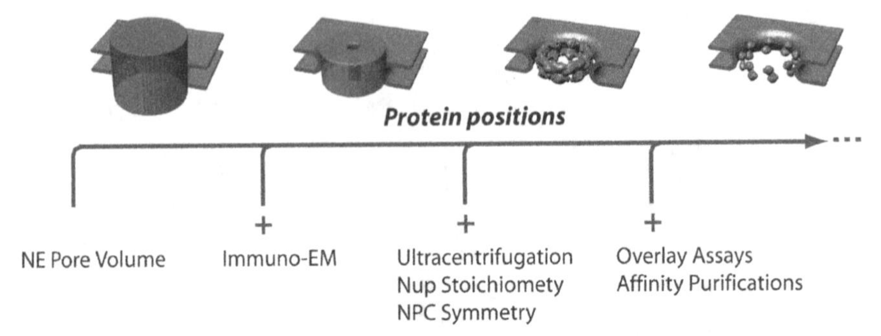

Fig. 6.7 Synergy between varied datasets results into increased precision of structure determination (Alber et al. 2007a; Alber et al. 2008; Robinson et al. 2007). The proteins are increasingly localized by the addition of different types of synergistic experimental information. As an example, each panel illustrates the localization of 16 copies of Nup192 in the ensemble of NPC structures generated, using the datasets indicated below. The smaller the volume (*red*), the better localized is the protein. The NPC structure is therefore essentially "molded" into shape by the large amount of experimental data

6.5 Conclusions

There is a wide spectrum of experimental and computational methods for identification and structural characterization of macromolecular complexes. The data from these methods need to be combined through integrative computational approaches to achieve higher resolution, accuracy, precision, completeness, and efficiency than any of the individual methods. New methods must be capable of generating possible alternative models consistent with information such as stoichiometry, interaction data, similarity to known structures, docking results, and low-resolution images.

Structural biology is a great unifying discipline of biology. Thus, structural characterization of many protein complexes will bridge the gaps between genome sequencing, functional genomics, proteomics, and systems biology. The goal seems daunting, but the prize will be commensurate with the effort invested, given the importance of molecular machines and functional networks in biology and medicine.

Acknowledgments We are grateful to Wah Chiu, David Agard, Wolfgang Baumeister, Joachim Frank, Fred Davis, M.S. Madhusudan, Min-yi Shen, Keren Lasker, Daniel Russell, Friedrich Foerster, Dmitry Korkin, Maya Topf and Ben Webb for many discussions about structure characterization by satisfaction of spatial restraints. We are also thankful to Svetlana Dokdovskaya, Liesbeth Veenhoff, Whenzu Zhang, Julia Kipper, Damien Devos, Adisetyantari Suprapto, Orit Karni-Schmidt, and Rosemary Williams for their contribution to the determination of the NPC structure. We acknowledge support from the Sandler Family Supporting Foundation, NIH/NCRR U54 RR022220, NIH R01 GM54762, Human Frontier Science Program, NSF IIS 0705196, and NSF EIA-0324645. And we are grateful for computer hardware gifts from Ron Conway, Mike Homer, Intel, Hewlett-Packard, IBM, and Netapp. This review is based on refs. (Alber et al. 2007a; Alber et al. 2007b; Alber et al. 2008; Robinson et al. 2007).

References

Abbott, A. (2002). The society of proteins. Nature, *417*(6892), 894–896.

Aebersold, R. and Mann, M. (2003). Mass spectrometry-based proteomics. Nature, *422*(6928), 198–207.

Alber, F., Dokudovskaya, S., Veenhoff, L., Zhang, W., Kipper, J., Devos, D., Suprapto, A., Karni-Schmidt, O., Williams, R., Chait, B. T., Rout, M. P. and Sali, A. (2007a). Determining the architectures of macromolecular Assemblies. Nature, *450*(7170), 683–694.

Alber, F., Dokudovskaya, S., Veenhoff, L., Zhang, W., Kipper, J., Devos, D., Suprapto, A., Karni-Schmidt, O., Williams, R., Chait, B. T., Sali, A. and Rout, M. P. (2007b). The molecular architecture of the nuclear pore complex. Nature, *450*(7170), 695–701.

Alber, F., Eswar, N. and Sali, A. (2004). Structure determination of macromolecular complexes by experiment and computation. In J. M. Bujnicki (Ed.), *Practical Bioinformatics* (pp. 73–96). Germany: Springer-Verlag.

Alber, F., Foerster, F., Korkin, D., Topf, M. and Sali, A. (2008). Integrating diverse data for structure determination of macromolecular assemblies. Ann Rev Biochem, *77*, in press.

Alber, F., Kim, M. F. and Sali, A. (2005). Structural characterization of assemblies from overall shape and subcomplex compositions. Structure, *13*(3), 435–445.

Alberts, B. (1998). The cell as a collection of protein machines: preparing the next generation of molecular biologists. Cell, *92*(3), 291–294.

Bauer, A. and Kuster, B. (2003). Affinity purification-mass spectrometry. Powerful tools for the characterization of protein complexes. Eur J Biochem, *270*(4), 570–578.

Brunger, A. T. (1993). Assessment of phase accuracy by cross validation: the free R value. Methods and applications. Acta Crystallogr D Biol Crystallogr, *49*(Pt 1), 24–36.

Collins, S. R., Kemmeren, P., Zhao, X. C., Greenblatt, J. F., Spencer, F., Holstege, F. C., Weissman, J. S. and Krogan, N. J. (2007). Toward a comprehensive atlas of the physical interactome of Saccharomyces cerevisiae. Mol Cell Proteomics, *6*(3), 439–450.

Davis, F. P. and Sali, A. (2005). PIBASE: a comprehensive database of structurally defined protein interfaces. Bioinformatics, *21*(9), 1901–1907.

Devos, D., Dokudovskaya, S., Williams, R., Alber, F., Eswar, N., Chait, B. T., Rout, M. P. and Sali, A. (2006). Simple fold composition and modular architecture of the nuclear pore complex. Proc Natl Acad Sci U S A, *103*(7), 2172–2177.

Devos, D., Dokudovskaya, S., Williams, S., Alber, F., Williams, R., Chait, B. T., Rout, M. P. Sali, A. (2004) Components of coated vesicles and nuclear pore complexes share a commmon molecular architecture. PL.S Biol,*2*(12), e380.

Devos, D. and Russell, R. B. (2007). A more complete, complexed and structured interactome. Curr Opin Struct Biol, *17*(3), 370–377.

Fiaux, J., Bertelsen, E. B., Horwich, A. L. and Wuthrich, K. (2002). NMR analysis of a 900 K GroEL GroES complex. Nature, *418*(6894), 207–211.

Frank, J. (2006). *Three-Dimensional Electron Microscopy of Macromolecular Assemblies*. Oxford: Oxford University Press.

Gavin, A. C., Aloy, P., Grandi, P., Krause, R., Boesche, M., Marzioch, M., Rau, C., Jensen, L. J., Bastuck, S., Dumpelfeld, B., Edelmann, A., Heurtier, M. A., Hoffman, V., Hoefert, C., Klein, K., Hudak, M., Michon, A. M., Schelder, M., Schirle, M., Remor, M., Rudi, T., Hooper, S., Bauer, A., Bouwmeester, T., Casari, G., Drewes, G., Neubauer, G., Rick, J. M., Kuster, B., Bork, P., Russell, R. B. and Superti-Furga, G. (2006). Proteome survey reveals modularity of the yeast cell machinery. Nature, *440*(7084), 631–636.

Harris, M. E., Nolan, J. M., Malhotra, A., Brown, J. W., Harvey, S. C. and Pace, N. R. (1994). Use of photoaffinity crosslinking and molecular modeling to analyze the global architecture of ribonuclease P RNA. Embo J, *13*(17), 3953–3963.

Ito, T., Tashiro, K., Muta, S., Ozawa, R., Chiba, T., Nishizawa, M., Yamamoto, K., Kuhara, S. and Sakaki, Y. (2000). Toward a protein-protein interaction map of the budding yeast:

A comprehensive system to examine two-hybrid interactions in all possible combinations between the yeast proteins. Proc Natl Acad Sci U S A, *97*(3), 1143–1147.

Krogan, N. J., Cagney, G., Yu, H., Zhong, G., Guo, X., Ignatchenko, A., Li, J., Pu, S., Datta, N., Tikuisis, A. P., Punna, T., Peregrin-Alvarez, J. M., Shales, M., Zhang, X., Davey, M., Robinson, M. D., Paccanaro, A., Bray, J. E., Sheung, A., Beattie, B., Richards, D. P., Canadien, V., Lalev, A., Mena, F., Wong, P., Starostine, A., Canete, M. M., Vlasblom, J., Wu, S., Orsi, C., Collins, S. R., Chandran, S., Haw, R., Rilstone, J. J., Gandi, K., Thompson, N. J., Musso, G., St Onge, P., Ghanny, S., Lam, M. H., Butland, G., Altaf-Ul, A. M., Kanaya, S., Shilatifard, A., O'Shea, E., Weissman, J. S., Ingles, C. J., Hughes, T. R., Parkinson, J., Gerstein, M., Wodak, S. J., Emili, A. and Greenblatt, J. F. (2006). Global landscape of protein complexes in the yeast Saccharomyces cerevisiae. Nature, *440*(7084), 637–643.

Lakey, J. H. and Raggett, E. M. (1998). Measuring protein-protein interactions. Curr Opin Struct Biol, *8*(1), 119–123.

Lim, R. Y. and Fahrenkrog, B. (2006). The nuclear pore complex up close. Curr Opin Cell Biol, *18*(3), 342–347.

Malhotra, A. and Harvey, S. C. (1994). A quantitative model of the Escherichia coli 16 S RNA in the 30 S ribosomal subunit. J Mol Biol, *240*(4), 308–340.

Mendez, R., Leplae, R., Lensink, M. F. and Wodak, S. J. (2005). Assessment of CAPRI predictions in rounds 3-5 shows progress in docking procedures. Proteins, *60*(2), 150–169.

Phizicky, E., Bastiaens, P. I., Zhu, H., Snyder, M. and Fields, S. (2003). Protein analysis on a proteomic scale. Nature, *422*(6928), 208–215.

Robinson, C., Sali, A. and Baumeister, W. (2007). The molecular sociology of the cell. Nature, *450*(7172), 973–982.

Russell, R. B., Alber, F., Aloy, P., Davis, F. P., Korkin, D., Pichaud, M., Topf, M. and Sali, A. (2004). A structural perspective on protein-protein interactions. Curr Opin Struct Biol, *14*(3), 313–324.

Sali, A. (2003). NIH workshop on structural proteomics of biological complexes. Structure, *11*(9), 1043–1047.

Sali, A., Glaeser, R., Earnest, T. and Baumeister, W. (2003). From words to literature in structural proteomics. Nature, *422*(6928), 216–225.

Sali, A. and Kuriyan, J. (1999). Challenges at the frontiers of structural biology. Trends Cell Biol, *9*(12), M20–24.

Shen, M. Y. and Sali, A. (2006). Statistical potential for assessment and prediction of protein structures. Protein Sci, *15*(11), 2507–2524.

Trester-Zedlitz, M., Kamada, K., Burley, S. K., Fenyo, D., Chait, B. T. and Muir, T. W. (2003). A modular cross-linking approach for exploring protein interactions. J Am Chem Soc, *125*(9), 2416–2425.

Uetz, P., Giot, L., Cagney, G., Mansfield, T. A., Judson, R. S., Knight, J. R., Lockshon, D., Narayan, V., Srinivasan, M., Pochart, P., Qureshi-Emili, A., Li, Y., Godwin, B., Conover, D., Kalbfleisch, T., Vijayadamodar, G., Yang, M., Johnston, M., Fields, S. and Rothberg, J. M. (2000). A comprehensive analysis of protein-protein interactions in Saccharomyces cerevisiae. Nature, *403*(6770), 623–627.

Valencia, A. and Pazos, F. (2002). Computational methods for the prediction of protein interactions. Curr Opin Struct Biol, *12*(3), 368–373.

Yang, Q., Rout, M. P. and Akey, C. W. (1998). Three-dimensional architecture of the isolated yeast nuclear pore complex: functional and evolutionary implications. Mol Cell, *1*(2), 223–234.

Chapter 7
Topological and Dynamical Properties of Protein Interaction Networks

Sergei Maslov

Abstract This chapter reviews some of the recent research on topological and dynamical properties of Protein-protein Interaction (PPI) networks. In its first part we describe the set of numerical algorithms aimed at: 1) constructing a null-model random network with a desired set of low-level topological properties; 2) detection of over- or under-represented topological patterns such as degree-degree correlations between interacting nodes. In the second part of the chapter we describe a recently developed set of computational tools and analytical methods which allow one to go beyond purely topological studies of PPI networks and efficiently calculate the mass-action equilibrium of protein concentrations and its response to systematic perturbations. In particular, we explore how large (several-fold) changes in total abundance of a small number of proteins shift the equilibrium between free and bound concentrations of proteins throughout the PPI network. Our primary conclusion is that, on average, the effects of such perturbations exponentially decay with the network distance away from the perturbed node. This explains why, despite globally connected topology, individual functional modules in such networks are able to operate fairly independently. Under specific favorable conditions, realized in a significant number of paths in the yeast PPI network, concentration perturbations can selectively propagate over considerable network distances (up to four steps). Such "action-at-a-distance" requires high concentrations of heterodimers along the path as well as low free (unbound) concentration of intermediate proteins.

7.1 Introduction

Ever since the appearance of the first experiments detecting protein-protein binding interactions on an organism-wide scale the topology of the resulting Protein-protein Interaction (PPI) networks has attracted a lot of attention. Now that

---

S. Maslov
Brookhaven National Laboratory, Department of Condensed Matter Physics and Materials Science, Upton, New York, USA
e-mail: maslov@bnl.gov

A. Panchenko, T. Przytycka (eds.), *Protein-protein Interactions and Networks*,
DOI: 10.1007/978-1-84800-125-1_7,

genome-wide (or nearly genome-wide) networks are known for 6 model organisms (and still counting) [1,2,3,4,5,6,7,8,9,10,11,12,13] one could start making some general observations about their unusual large-scale topological properties. They are:

- The broad distribution of the number of binding partners of individual proteins (their network degrees) [14,15]. The exact functional form of the degree histogram remains the subject of a heated discussion but I think that most parties would agree that it is unusually broad. Hence the existence of hubs - proteins having a disproportionately large number (from tens to hundreds) of direct binding partners.
- Anti-correlation between network degrees of interacting proteins [16]. In such network architecture hub proteins avoid directly or indirectly linking to each other and instead tend to interact with proteins of low connectivity/degree.
- The "small world" effect in which most pairs of protein nodes (about 80% for most networks) are linked to each other [17] by a relatively short chain of interactions involving several intermediate proteins. This property is nearly inevitable consequence of the existence of hubs and a broad degree distribution. While facilitating meaningful signaling it also presents a potential problem by providing a conduit for propagation of undesirable cross-talk between individual functional units/pathways. Later on in this chapter we will attempt to quantify the magnitude of cross-talk mediated by PPI networks and point out their topological and dynamical properties that reduce the severity of this problem.
- The existence of densely interconnected modules (or clusters or communities) correlated with biological function and large multi-protein complexes [18].

Protein-protein binding interactions are naturally described in terms of a weighted network in which individual edges are graded by their binding strength. Therefore, while purely topological (binary) analysis of these networks in itself constitutes a fascinating subject, ultimately, it constitutes just the first small step towards addressing other biologically important questions as:

- The evolutionary history of PPI networks and dominant processes modifying its nodes and the binding strength of its edges on an evolutionary timescale.
- Steady state and dynamical properties of a binding equilibrium state determined by protein concentrations and the set of dissociations constants of individual protein-protein interactions.
- Robustness and stability of this equilibrium state with respect to noise and perturbations.

In the second part of this chapter I will move beyond purely topological analysis to quantify the binding equilibrium in graded PPI networks and thus touch upon the last two topics from the list.

7.2 Detecting Non-Random Topological Patterns in PPI Networks

PPI networks (as any other biological entities) lack the top-down design. Instead, selective forces of biological evolution shape them from raw material provided by random events such as mutations within individual genes, and gene duplications. As a result their connections are characterized by a large degree of randomness. One may wonder which connectivity patterns are indeed random, and which arose due to particular mechanisms of the network growth, evolution, and/or its functional design principles and limitations? In this chapter I will describe a general set of algorithms aimed at detection of topological patterns in these networks that strongly deviate from random null-model expectations and thus are of a potential functional and/or evolutionary significance.

7.2.1 Single-Node Topological Properties: Degree Distribution

The first markedly non-random feature of PPI networks is an extremely broad distribution of nodes' degrees (sometimes also called connectivities) defined as the number of immediate binding partners of a given protein [14,15]. While the majority of proteins have just a few binding interactions with other proteins, there exist some protein nodes, to which we will refer to as "hubs", with an unusually large number of immediate binding partners. Degrees of the most connected nodes in such a network is typically several orders of magnitude larger than the average degree of a node in the network. Often the degree histogram $N(K)$ can be approximated by a power law $N(K) \sim K^{-\gamma}$ in which case the network is referred to as scale-free [14]. The empirically observed values of the exponent γ typically range between 2 and 3.

High throughput Yeast Two-Hybrid (Y2H) experiments consistently generate networks with scale-free or nearly scale-free architecture (the corresponding plots are all presented e.g. in Ref. [19]). Another high-throughput experimental technique - tandem affinity purification-mass spectrometry (TAP-MS) - also generates networks with a broad degree distribution, which however is markedly different from a simple power-law functional form possibly due to inclusion of indirect interactions [20] (that is to say interaction links between proteins in the same multi-protein complex that are not in direct contact with each other).

Three explanation/simple models interpreting the ubiquitous broad degree distribution in PPI networks were proposed in the recent literature:

- The first, *evolutionary*, explanation relies on duplication-divergence models [21,22,23,20] to explain the appearance of protein hubs. The linear preferential attachment term, which is known to generate networks with scale-free topology [14], naturally appears in these models for the following reason: a protein with K binding partners has K chances to get a new partner as a result of duplication of already existing one. In case of a growing genome this gives rise to a broad degree distribution consistent with that measured in PPI networks. With small modifications these models are also applicable to explain the broad degree distribution in any protein network.

- The second, *biophysical*, explanation, first proposed in [24] and later refined and extended in [25,19], does not rely on the assumption of an expanding genome. Instead, it explains the appearance of the hub proteins by variations in proteins' overall "stickiness" quantified by the average hydrophobicity of their surfaces. The likelihood of detection of a given protein-protein binding interaction depends on its dissociation constant K_{ij} which is proportional to the *exponent* of binding free energy. Thus a narrow (Gaussian) distribution of the latter would corresponds to a broad (log-normal) distribution of the former, which in its turn could result in a broad degree distribution. Biophysical models [24,25,19] assume that the binding free energy of a pair of interacting proteins is linearly correlated with the *sum* of their individual stickiness factors. This assumption turns out to be sufficient to give rise to PPI networks with realistic topological properties. Unlike duplication-divergence models this explanation is uniquely applicable to PPI networks and sheds no light on broad degree distribution observed in other biological networks.
- Finally, the third, *functional*, explanation interprets the broad degree distribution in terms of large variability in the complexity of tasks performed by an organism. Proteins participating in simpler tasks may need just a few interaction partners, while those used in more complicated and global tasks become hubs. An evidence in favor of this interpretation is provided by the positive correlation [15] between protein's degree and its essentiality (lethality of its null mutant). This functional explanation of the broad degree distribution doesn't exclude the previous two. In fact it relies on gene duplications and non-specific "sticky" interactions between proteins to generate the new binding pairs from which the evolution would subsequently select the functionally important ones.

One aspect of a broad distribution of node degrees in protein interaction networks, is the possibility of amplification and exponential spread of signals and perturbations propagating in the network. The upper bound on the one step amplification of a signal or perturbation propagating in an undirected network is given by

$$\mathcal{A} = \frac{\langle K(K-1)\rangle}{\langle K\rangle}. \tag{7.1}$$

Here and in what follows the angular brackets $\langle\cdot\rangle$ denote the averaging over all nodes in the network. Thus defined amplification factor $\mathcal{A}$ measures the average number of neighbors to which the signal/perturbation can be potentially transmitted in one step. The above equation follows from the observation that a signal/perturbation arrives at a given node via one of its K edges and leaves along any of its $K-1$ remaining links. Hence the probability for a node to be involved in this process linearly scales with its degree K favoring hubs over low-connectivity nodes. The above picture might be generalized if one assumes a certain finite probability p for a perturbation to spread to a given neighbor. For $p\mathcal{A} \leq 1$ perturbations eventually dies out and hence affects only a small fraction of nodes in the network. On the other hand, for $p\mathcal{A} > 1$ perturbations propagating in the network might be exponentially amplified, and thus interfere with functional signals over the entire network.

The amplification factor $\mathcal{A}$ in scale-free networks is typically very large. For example, its value calculated in the yeast PPI network based on the full dataset of Ref. [2] is as high as 26. This is because for $\gamma < 3$ the sum $\sum_{K=1}^{K_{max}} K(K-1) K^{-\gamma}$ defining the network average $\langle K(K-1)\rangle$ and approximated by an integral $\int_1^{K_{max}} K(K-1)K^{-\gamma} dK$ formally diverges at its upper cutoff. In practice that means that the value of $\mathcal{A}$ is determined by degrees K_{max} of the few largest hubs in the system. This is why undesirable cross-talk propagation through a network of physical interactions presents a potential problem. Later on we would return to the question of cross-talk and demonstrate how both the topological [16] and the equilibrium/dynamical [26] properties of PPI networks help to reduce such undesirable interference between functional biological signals.

7.2.2 Edge Swapping Algorithm: Constructing a Randomized Network

The first step in detecting non-random topological patterns in a network is to construct its meaningful randomized version (the null-model). A broad distribution of degrees indicates that the degree itself is an important characteristic of a node and as such it should be preserved in its randomized version [16]. Higher-level topological patterns in such networks could only be discovered by comparison with the null-model that has the same degree distribution as the original network. A simple yet efficient algorithm generating a random network (or, if desired, multiple realizations of random networks) that preserves degrees of individual nodes was proposed in [27,16]. It consists of multiple repetitions of the following simple edge swapping move (elementary rewiring step) illustrated in Fig. 7.1:

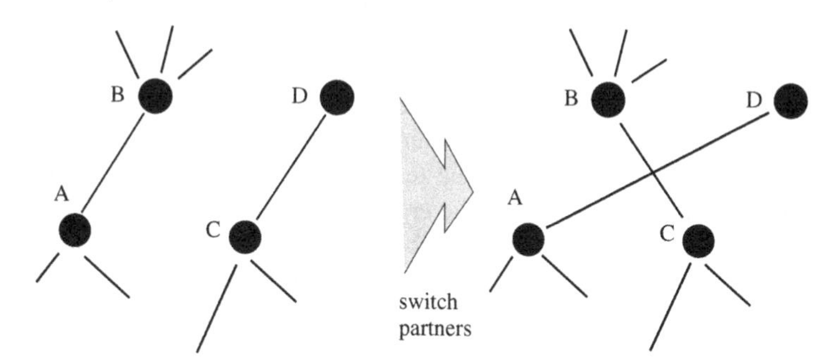

Fig. 7.1 One step of the random local rewiring algorithm. A pair of edges A–B and C–D is randomly selected. The two edges are then rewired in such a way that A becomes connected to D, while C to B, provided that none of these new edges already exist in the network, in which case the rewiring step is aborted and a new pair of edges is selected. An independent random network is obtained when the above local switch move is performed a large number of times (several times in excess of the total number of edges in the network)

Randomly select a pair of edges A–B and C–D and rewire them in such a way that A becomes connected to D, while C to B. To prevent the appearance of multiple edges connecting the same pair of nodes, the rewiring step is aborted and a new pair of edges is selected if any one of the new edges already exists in the network. A repeated application of the above rewiring step leads to a randomized version of the original network. A rule of the thumb we use in most of our simulations is that in a network containing E edges the edge-swapping step needs to be repeated around $3E$ time to generate a fully-randomized null-model network. The set of MATLAB programs generating such a randomized version of any complex network can be downloaded from [28].

Sometimes it is desirable that a null-model random network in addition to nodes' degrees conserves some other quantitative topological property of a network. In this case the random rewiring algorithm described above should be supplemented with the additional Metropolis acceptance/rejection criterion [29]. Let's consider a concrete example in which one wants to generate a random network with the same set of nodes' degrees and the same number $N^{(\Delta)}$ of triangles as in the original undirected network [29]. The number of triangles in a network is closely related to its "clustering coefficient" which is a measure of network's modularity [30]. Hence, by conserving $N^{(\Delta)}$ one generates a null-model with the same average level of modularity as the original network. The Metropolis version [29] of the random rewiring algorithm uses an auxiliary "energy function" H that favors the number of triangles in a randomized network $N_r^{(\Delta)}$ to be as close as possible to its value $N^{(\Delta)}$ in the original network:

$$H = \frac{\left(N_r^{(\Delta)} - N^{(\Delta)}\right)^2}{N^{(\Delta)}} \tag{7.2}$$

As usual the Metropolis rules accept any edge-swapping step that lowers the energy H or leaves it unchanged. On the other hand, steps leading to a ΔH increase in the "energy" H are accepted with probability $\exp(-\Delta H/T)$. Here the exact rules of the algorithm depend on (typically very small) "temperature" T introduced to prevent the sequence of rewiring steps from getting stuck in a local (often non-representative) energy minimum. In order to get a random network with $N_r^{(\Delta)}$ sufficiently close to $N^{(\Delta)}$ the temperature should be selected to be as small as possible without sacrificing the ergodicity of the problem. In the end one could always "prune" the resulting ensemble of random networks by leaving only networks with $N_r^{(\Delta)} = N^{(\Delta)}$. Another option is to perform a simulated annealing randomization scheme that starts with a relatively high temperature and then gradually lowers it to zero.

7.2.3 Detecting Non-Random Topological Patterns in a Network

Once the desired null model randomized network is generated one could use it to find out which topological quantities in the real complex network significantly deviate from their values in this null model. Such deviations are best quantified by the following two ratios.

The first ratio is

$$R(j) = \frac{N(j)}{N_r(j)} \tag{7.3}$$

where $N(j)$ is the number of times the pattern j is seen in the real network, and $N_r(j)$ is the average number of its occurrences in an ensemble of randomized networks, generated e.g. by one of the local rewiring algorithms described above. Patterns selected by design or evolution of the network would manifest themselves by $R(j) > 1$, while suppressed patterns correspond to $R(j) < 1$.

While $R(j)$ determines the magnitude of the suppression/enhancement it tells one nothing about the statistical significance of the effect. This latter quantity is quantified by another ration – the Z-score of the deviation:

$$Z(j) = \frac{N(j) - N_r(j)}{\Delta N_r(j)}, \tag{7.4}$$

where $\Delta N_r(j)$ is the standard deviation of $N_r(j)$ measured in a sufficiently large ensemble of randomized networks.

Alternatively the statistical significance of the difference between real and randomized networks can be quantified in terms of its P-value. The P-value is defined as the probability that the number of patterns $N_r(j)$ in a randomized network is larger or equal (or smaller or equal in case when $N(j) < N_r(j)$) than $N(j)$. If one can verify that $N_r(j)$ is a Gaussian-distributed random variable, the Z-score can be easily converted to the P-value.

One particular case of detection of non-random topological patterns in networks was presented in [31,32]. The authors of this study first exhaustively labeled all three- and four-node subgraphs such as e.g. a feed-forward or a feedback loops in directed networks. Then they identified the network motifs j whose abundance $N(j)$ in the real complex network is significantly higher (or lower) than null-model expectations in a random network. To properly identify over- or under-represented higher-order (e.g. four-node) motifs one needs to factor in already detected non-random patterns on a three-node level. To achieve this goal the Ref. [32] used our edge-swapping Metropolis algorithm [29] (see Eq. 7.2) preserving all statistically significant three-node motifs. The number $N(j)$ of four-node motifs contained in a given complex network was then compared to its expected value in a randomized network preserving not only the in- and out-degrees but also the numbers of all significantly over- or under-represented three-node subgraphs.

7.2.4 An Example: Correlations Between Degrees of Neighboring Nodes

This section presents another example of applying the general pattern-detection scheme described in the previous section.

The *correlation profile* of a network quantifies correlations between degrees of its neighboring nodes [16,29]. To find out the exact pattern of these correlations one compares $N(K_0, K_1)$ – the number of edges connecting nodes of degree K_0 to those of degree K_1 – and $N_r(K_0, K_1) \pm \Delta N_r(K_0, K_1)$, which is its value in a randomized network generated by the edge-swapping algorithm. Correlations manifest themselves as systematic deviations of the ratio

$$R(K_0, K_1) = N(K_0, K_1)/N_r(K_0, K_1) \qquad (7.5)$$

away from 1. Statistical significance of such deviations is quantified by their Z-score

$$Z(K_0, K_1) = (N(K_0, K_1) - N_r(K_0, K_1))/\Delta N_r(K_0, K_1)\,, \qquad (7.6)$$

where $\Delta N_r(K_0, K_1)/N$ is the standard deviation of $N_r(K_0, K_1)$ in an ensemble of randomized networks.

Multiple randomized versions of a protein binding network obtained in a high throughput Y2H experiment of Ito et al. [2] were constructed by randomly rewiring its edges, while preventing multiple connections between a given pair of nodes, as described in the previous chapter.

Figure 7.2 shows the ratio $R(K_0, K_1)$ while Fig. 7.3 – the statistical significance $Z(K_0, K_1)$ of deviations between the real and randomized networks visible in Fig. 7.2. To calculate these ratios 1000 randomized networks were sampled and degrees were logarithmically binned into two bins per decade. The combination of R- and Z-profiles reveals the regions on the $K_0 - K_1$ plane, where connections between proteins in the real network are significantly enhanced or suppressed, compared to the null model. In particular, the dark region in the upper right corner of Figs. 7.2, 7.3 reflects the reduced likelihood that two hubs are directly linked to each other, while light-colored regions in the upper left and the lower right corners of these figures reflect the tendency of hubs to associate with nodes of low degree. One should also note a prominent feature on the diagonal corresponding to an enhanced affinity of proteins with between 4 and 9 physical interaction partners towards each other. This feature can be tentatively attributed to members of multi-protein complexes interacting with other proteins from the same complex. The above range of degrees thus correspond to a typical number of direct interaction partners of a protein in a multi-protein complex. When we studied pairs of interacting proteins in this range of degrees we found 39 of such pairs to belong to the same

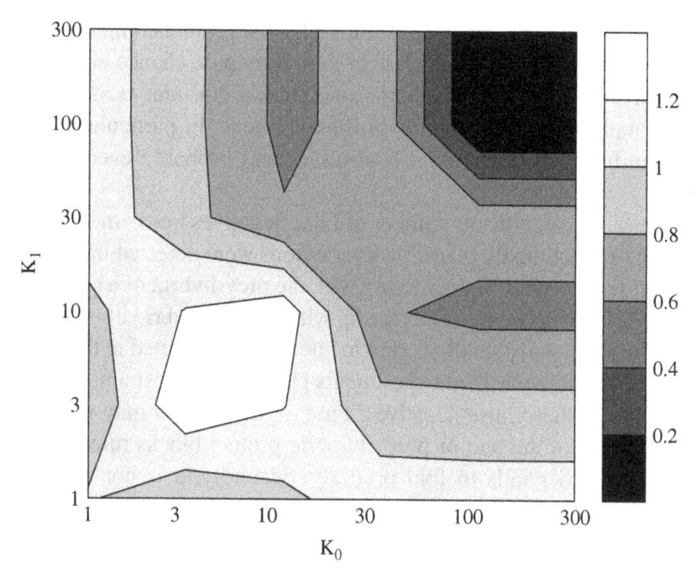

Fig. 7.2 Correlation profile of the protein interaction network in yeast. The ratio $R(K_0, K_1) = N(K_0, K_1)/N_r(K_0, K_1)$. Here $N(K_0, K_1)$ is the number of edges connecting nodes of degree K_0 to those of degree K_1 in the full yeast PPI set of Ref. [2], while $N_r(K_0, K_1)$ is the same quantity in a randomized version of the same network, generated by the random rewiring algorithm described in the text. Note the logarithmic scale of both axes

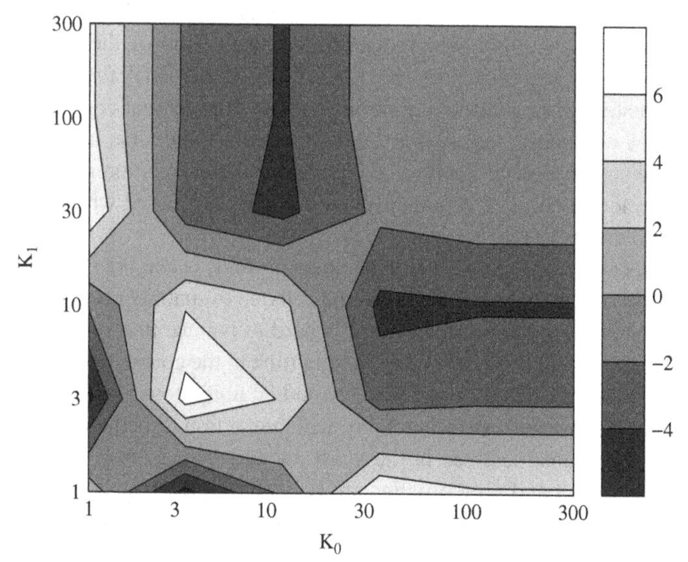

Fig. 7.3 Statistical significance of correlations present in the protein interaction network in yeast quantified by the Z-score (Eq. 7.6). Here $N(K_0, K_1)$ is the number of edges connecting nodes of degree K_0 to those of degree K_1 in the full yeast PPI set of Ref. [2], while $N_r(K_0, K_1)$ is the same quantity in a randomized version of the same network, generated by the random rewiring algorithm described in the text, and $\Delta N_r(K_0, K_1)$ is the standard deviation of $N_r(K_0, K_1)$ measured in 1000 realizations of the randomized network. Note the logarithmic scale of both axes

complex in the recent high-throughput study of yeast protein complexes [9]. This is about 4 times more than one would expect to find by pure chance alone.

When analyzing molecular networks one should consider possible sources of errors in the underlying data. Two-hybrid experiments in particular are known to contain a significant number of false positives and probably even more of false negatives.

The evidence of a significant number of false negatives lies in the fact that only a small fraction of functionally plausible interactions were detected in both directions (the bait-hybrid of a protein A interacting with the prey-hybrid of a protein B as well as the prey-hybrid of a protein A interacting with the bait-hybrid of a protein B). It is also attested by a relatively small overlap in interactions detected in the two independent high-throughput two hybrid experiments [1,2]. There exist a number of plausible explanations of these false negatives. First of all, binding may not be observed if the conformation of the bait or prey chimeric protein blocks relevant interaction sites or if it altogether fails to fold properly. Secondly, it is not entirely clear if the number of cells in batches used in high-throughput two hybrid experiments is sufficient for any given bait-prey pair to meet in at least one cell. Finally, 391 out of potential 5671 baits in Ref. [2] were not experimentally tested because they were found to activate the transcription of the reporter gene in the absence of any prey proteins.

Several sources of false positives are also commonly mentioned in the literature:

- In one scenario spurious interactions of highly connected baits are thought to arise due to a *low-frequency* indiscriminate activation of the reporter gene in the absence of any prey proteins. Such false positives (if they exist) are easy to eliminate by using curated high-throughput datasets that contain only protein pairs that were observed, say, at least 3 times in the course of the experiment. We have shown that all qualitative features of the correlation profile of the protein interaction network reported above remain unchanged when one uses such curated datasets [33].
- In another scenario the interaction between proteins is real but it never happens in the course of the normal life cycle of the cell due to spatial or temporal separation of participating proteins. However, it is hard to believe that such non-functional interactions would be preserved for a long time in the course of evolution. Hence, it is dubious that such false-positives would be ubiquitous.
- In yet another scenario an indirect physical interaction is mediated by one or more unknown proteins localized in the yeast nucleus. However, since in two-hybrid experiments bait and prey proteins are typically highly overexpressed, it is only very abundant intermediate proteins that can give rise to an indirect binding. The relative insignificance of indirect bindings is attested by a relatively small number of triangles (178 vs $\propto$ 100 in a randomized version) in the protein interaction network. Indeed, an indirect interaction of a protein A with a protein B effectively closes the triangle of direct interactions A-C and C-B with an intermediate protein C.

Fortunately, qualitative features of correlation profiles of complex networks are very robust with respect to unbiased set of false positives and false negatives. Indeed, as previously undetected edges are added to the network (or falsely detected edges are removed from it) the average connectivity of its nodes changes. As a result correlation features visible in its correlation profiles may shift their positions and intensity, but are likely to preserve their qualitative characteristics up to a very high level of false positives or false negatives.

The data for the yeast PPI network analyzed in this section come from a high throughput experiment performed in one lab using a unique experimental technique [2]. This fact makes it a perfect candidate for correlation profiling. Indeed, since almost all pairs of yeast proteins were tested as potential interacting partners, the statistical information contained in the resulting network contains no anthropogenic bias. On the other hand, when the information about edges in a network is obtained from a database, combining results of many experimental groups using various techniques, one should worry about a hidden anthropogenic factor: some proteins just constitute more attractive subjects of research and are, therefore, relatively better studied than the others. The level of clustering in networks based on the database data may be overestimated due to several reasons: 1) With the exception of systemwide experiments such as high-throughput two-hybrid screens in yeast [1,2], experimentalists are more likely to check for interactions between pairs of proteins within the same functional group. 2) A complete analysis of all possible pairwise interactions within a small group of proteins would influence the level of clustering in the network. In this case this group would manifest itself by relatively dense pattern of interactions with other members of the same group.

7.3 Equilibrium and Dynamical Properties of PPI Networks

While topological properties of PPI networks in themselves constitute a fascinating subject, they represents just a first step towards more quantitative understanding of network's equilibrium and dynamical properties. In this chapter we make the next logical step in this direction. To this end we calculate the binding equilibrium of the PPI network in baker's yeast *S. cerevisiae* and quantitatively study its response to various perturbations. For the most part this chapter follows our earlier publications [26,34].

Networks of protein-protein physical interactions (PPI) are known to be interconnected on a genome-wide scale. In such "small-world" PPI networks most pairs of nodes can be linked to each other by relatively short chains of interactions involving just a few intermediate proteins [17]. While globally connected architecture facilitates biological signaling and possibly ensures a robust functioning of the cell following a random failure of its components [35], it also presents a potential problem by providing a conduit for propagation of undesirable cross-talk between individual functional modules and pathways. Indeed, large (several-fold) changes in proteins' levels in the course of activation or repression of a certain functional module affect bound concentrations of their immediate interaction partners. These changes have a potential to cascade down a small-world PPI network affecting the equilibrium between bound and unbound concentrations of progressively more distant neighbors including those in other functional modules. Most often such indiscriminate propagation would represent an undesirable effect that has to be either tolerated or corrected by the cell. On the other hand, a controlled transduction of reversible concentration changes along specific chains of interacting proteins may be used for biologically meaningful signaling and regulation. A routine and well known example of such regulation is inactivation of a protein by sequestration with its strong binding partner.

Below we quantitatively investigate how large concentration changes propagate in the PPI network of yeast *S. cerevisiae*. We focus on the non-catalytic or reversible binding interactions whose equilibrium is governed by the Law of Mass Action (LMA) and do not consider irreversible, catalytic processes such as protein phosphorylation and dephosphorylation, proteolytic cleavage, etc. While such catalytic interactions constitute the most common and best studied mechanism of intracellular signaling, they represent only a rather small minority of all protein-protein physical interactions (for example, only ~5% links in the yeast PPI network used in our study involve a kinase). Furthermore, the balance between free and bound concentrations of proteins matters even for irreversible (catalytic) interactions. For example, the rate of a phosphorylation reaction depends on the availability of free kinases and substrate proteins that are both controlled by the LMA equilibrium calculated here. Thus perturbations of equilibrium concentrations considered in this study could be spread even further by other mechanisms such as transcriptional and translational regulation, and irreversible posttranslational protein modifications.

7.3.1 The Assignment of Dissociation Constants K_{ij}

To illustrate the general principles on a concrete example, below we use a highly curated genome-wide network of protein-protein physical interactions in yeast (*S. cerevisiae*) that, according to the BIOGRID database [36], were independently confirmed in at least two publications. The topological data are combined with a genome-wide dataset of protein abundances (or total = free+bound intracellular concentrations) in the log-phase growth in rich medium, measured by the TAP-tagged western blot technique [37]. Average protein concentrations in this dataset range between 50 and 1,000,000 molecules/cell with the median value around 3000 molecules/cell. After keeping only the interactions between proteins with known concentrations we were left with 4185 binding interactions among 1740 proteins.

The BIOGRID database [36] lists all interactions as pairwise and thus lacks information about multi-protein complexes larger than dimers. Thus in the main part of this study we consider only homo- and hetero-dimers and ignore the formation of higher-order complexes. We have previously demonstrated [26] that multi-protein complexes could be easily incorporated into our analysis.

Furthermore, it was shown that taking into account such complexes leaves our results virtually unchanged.

The state-of-the-art genome-wide PPI datasets lack information on dissociation constants K_{ij} of individual interactions. The only implicit assumption is that the binding is sufficiently strong to be detectable by a particular experimental technique (some tentative bounds on dissociation constants detectable by different techniques were recently reported in [38]). A rough estimate of the average binding strength in functional protein-protein interactions could be obtained from the PINT database [39]. This database contains about 400 experimentally measured dissociation constants between wildtype proteins from a variety of organisms. In agreement with predictions of Refs. [40,25] the histogram of these dissociation constants has an approximately log-normal shape. The average relevant for our calculations is that of the *association* constant $\langle 1/K_{ij} \rangle$ =1/(5nM). Common sense dictates that the dissociation constant of a functional binding between a pair of proteins should increase with their abundances. The majority of specific physical interactions between proteins are neither too weak (to ensure a considerable number of bound complexes) nor unnecessarily strong. Indeed, there is little evolutionary pressure towards increasing the binding strength between a pair of proteins beyond the point when both proteins (or at least the rate limiting one) spend most of their time in the bound state. The balance between these two opposing requirements is achieved by the value of dissociation constant K_{ij} equal to a fixed fraction of the largest of the two abundances C_i and C_j of interacting proteins. In our simulations we used $K_{ij} = \max(C_i, C_j)/20$ in which case the average association constant nicely agrees with its empirical value (1/(5nM)) observed in the PINT database [39]. In addition to this, perhaps, more realistic assignment of dissociation constants we also simulated PPI networks in which dissociation constants of all 4185 edges in our network are *equal to each other* and given by 1nM, 10nM, 100nM, and 1μM.

7.3.2 Concentration-Coupled Proteins

The Law of Mass Action (LMA) relates the free (unbound) concentration F_i of a protein to its total (bound and unbound) concentration C_i as

$$F_i = \frac{C_i}{1 + \sum_j F_j/K_{ij}} \tag{7.7}$$

Here the sum over j includes all specific binding partners of the protein i with free concentrations F_j and dissociation constants K_{ij}. The above equation follows from the Law of Mass Action equilibrium value of the bound (dimer) concentration $D_{ij} = F_i F_j / K_{ij}$ and the mass conservation $C_i = F_i + \sum_j D_{ij}$. While in the general case the set of N nonlinear equations (7.7) does not allow for an analytical solution for F_i, they are readily solved numerically e.g. by successive iterations starting from $F_i = C_i$. Our MATLAB program solving this equation can be downloaded at (28).

To investigate how large changes in abundances of individual protein affect the equilibrium throughout the PPI network we performed a systematic numerical study in which we recalculated the equilibrium free concentrations of all protein nodes following a twofold increase in the total concentration of just one of them: $C_i \to 2C_i$. This was repeated for the source of twofold perturbation spanning the set of all 1740 of proteins in our network The magnitude of the initial perturbation was selected to be representative of a typical shift in gene expression levels or protein abundances following a change in external or internal conditions. Thus here we simulate the propagation of functionally relevant changes in protein concentrations and not that of background stochastic fluctuations. A change in the free concentration F_j of a protein j was deemed to be significant if it exceeded the 20% level, which according to Ref. [41] is the average magnitude of cell-to-cell variability of protein abundances in yeast. We refer to such protein pairs $i \to j$ as *concentration-coupled*. The detection threshold could be raised simultaneously with the magnitude of the initial perturbation. For example, we found that the list of concentration-coupled pairs changes very little if instead of twofold (+100%) perturbation and the 20% detection threshold one applies a sixfold (+500%) initial perturbation and twofold (100%) detection threshold.

In general we found that lists of concentration-coupled proteins calculated for different assignments of dissociation constants strongly overlap with each other. For example, more than 80% of concentration-coupled pairs observed for the variable $K_{ij} = \max(C_i, C_j)/20$ assignment described above were also detected for the uniform assignment $K_{ij} = \text{const} = 10\text{nM}$. This relative robustness of our results allows us to use the latter conceptually simple way to assign the dissociation constants to illustrate our findings in the rest of this chapter.

7.3.3 Cascading Concentration Changes in PPI Networks

Our main observations are:

- *On average*, the magnitude of cascading changes in equilibrium free concentrations exponentially decays with the distance from the source of a perturbation. This explains why, despite a globally connected topology, individual modules in such networks are able to function fairly independently.
- Nevertheless, specific favorable conditions identified in our study cause perturbations to selectively affect proteins at considerable network distances (sometimes as far as four steps away from the source). This indicates that in general, such cascading changes *could not be neglected* when evaluating the consequences of systematic changes in protein levels, e.g. in response to environmental factors, or in gene knockout experiments. Conditions favorable for propagation of perturbations combine high yet monotonically decreasing concentrations of all heterodimers along the path with low free (unbound) concentrations of intermediate

proteins. While reversible protein binding links are symmetric, the propagation of concentration changes is usually asymmetric with the preferential direction pointing down the gradient in the total concentrations of proteins.

The results of our quantitative network-wide analysis of these effects are summarized in Fig. 7.4 and Table 7.1. From Fig. 7.4B one concludes that the fraction of proteins with significantly affected free concentrations rapidly (exponentially) decays with the length L of the shortest path (network distance) from the perturbed protein. The same statement holds true for bound concentrations if the distance is measured as the shortest path from the perturbed protein to any of the two proteins forming a heterodimer. Thus, on average, the propagation of concentration changes along the PPI network is indeed considerably dampened. On the other hand, from Table 7.1 one concludes that the total number of multi-step chains along which concentration changes propagate with little attenuation remains significant for all but the largest values of the dissociation constant. These two observations do not contradict each other since the number of proteins separated by distance L (the last column in Table 7.1) rapidly grows with L.

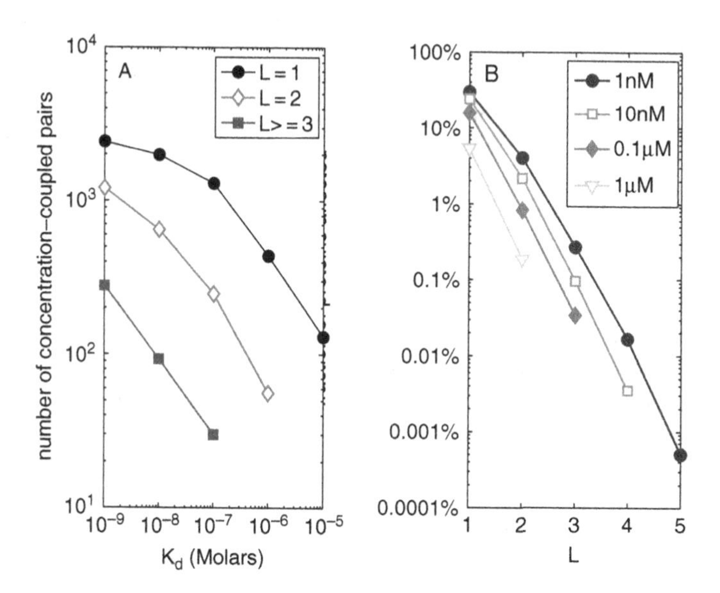

Fig. 7.4 (A) The statistics of propagation of concentration changes. The number of concentration coupled protein pairs versus the dissociation constant K_{ij} = const = K_d. Different curves correspond to different network distances L separating two proteins: $L = 1$ (*solid circles*), $L = 2$ (*empty diamonds*), and $L \geq 3$ (*solid squares*). Note that for large K_d the number of concentration-coupled pairs decays as $1/\sqrt{K_d}$. (B) Indiscriminate propagation of concentration perturbations is exponentially suppressed. The fraction of proteins with free concentrations affected by more than 20% among all proteins at network distance L from the perturbed protein. Different curves correspond to simulations with K_{ij} = const = 1nM (*solid circles*), 10nM (empty squares), 0.1μM (solid diamonds), and 1μM (*empty triangles*)

Table 7.1 The number of concentration-coupled pairs of yeast proteins separated by network distance L. Numerical simulations (twofold initial perturbation, 20% detection threshold) were performed for different assignment of dissociation constants: $K_{ij} = \max(C_i, C_j)/20$ (column 2), K_{ij} = const =1nM, 10nM, 0.1μM,1μM (columns 3–6). The column 7 lists the total number of protein pairs at distance L

L	var. 5nM	1nM	10nM	0.1μM	1μM	all
1	2003	2469	1915	1184	387	8168
2	415	1195	653	206	71	29880
3	15	159	49	8	0	87772
4	2	60	19	0	0	228026
5	0	3	0	0	0	396608

7.3.4 Conditions Favoring the Multi-Step Propagation of Perturbations

What conditions favor the multi-step propagation of perturbations along particular channels? In Fig. 7.5A we show a group of highly abundant proteins along with all binding interactions between them. Then on panel B of the same figure we show only those interactions that according to our LMA calculation give rise to highly abundant heterodimers (equilibrium concentration >1000 per cell). This breaks the densely interconnected subnetwork drawn in the panel A into 10 mutually isolated clusters. Some of these clusters contain pronounced linear chains that serve as conduits for propagation of concentration perturbations. The fact that perturbations indeed tend to propagate via highly abundant heterodimers is illustrated in the next panel (Fig. 7.5 C) where red arrows correspond to concentration-coupled nearest neighbors. Evidently, the edges in panels B and C largely (but not completely) coincide. Additionally, the panel C defines the preferred direction of propagation of perturbations from a more abundant protein to its less abundant binding partners. To further investigate what causes concentration changes to propagate along particular channels we took a closer look at eight three-step chains $A \to A_1 \to A_2 \to B$ with the largest magnitude of perturbation of the last protein B (twofold detection threshold following a twofold initial perturbation). The identification of intermediate proteins A_1 and A_2 was made by a simple optimization algorithm searching for the largest overall magnitude of intermediate perturbations along all possible paths connecting A and B.

Inspection of the parameters of these chains shown in Fig. 7.6 allows one to conjecture that for a successful transduction of concentration changes, the following conditions should be satisfied:

- Heterodimers along the whole path have to be of sufficiently high concentration D_{ij}.

Fig. 7.5 (continued) abundance A reduces free concentration of its immediate binding partner B by 20% or more. Note that links roughly coincide with highly abundant dimers shown in the panel B. Arrows reveal the preferential direction of propagation of perturbations

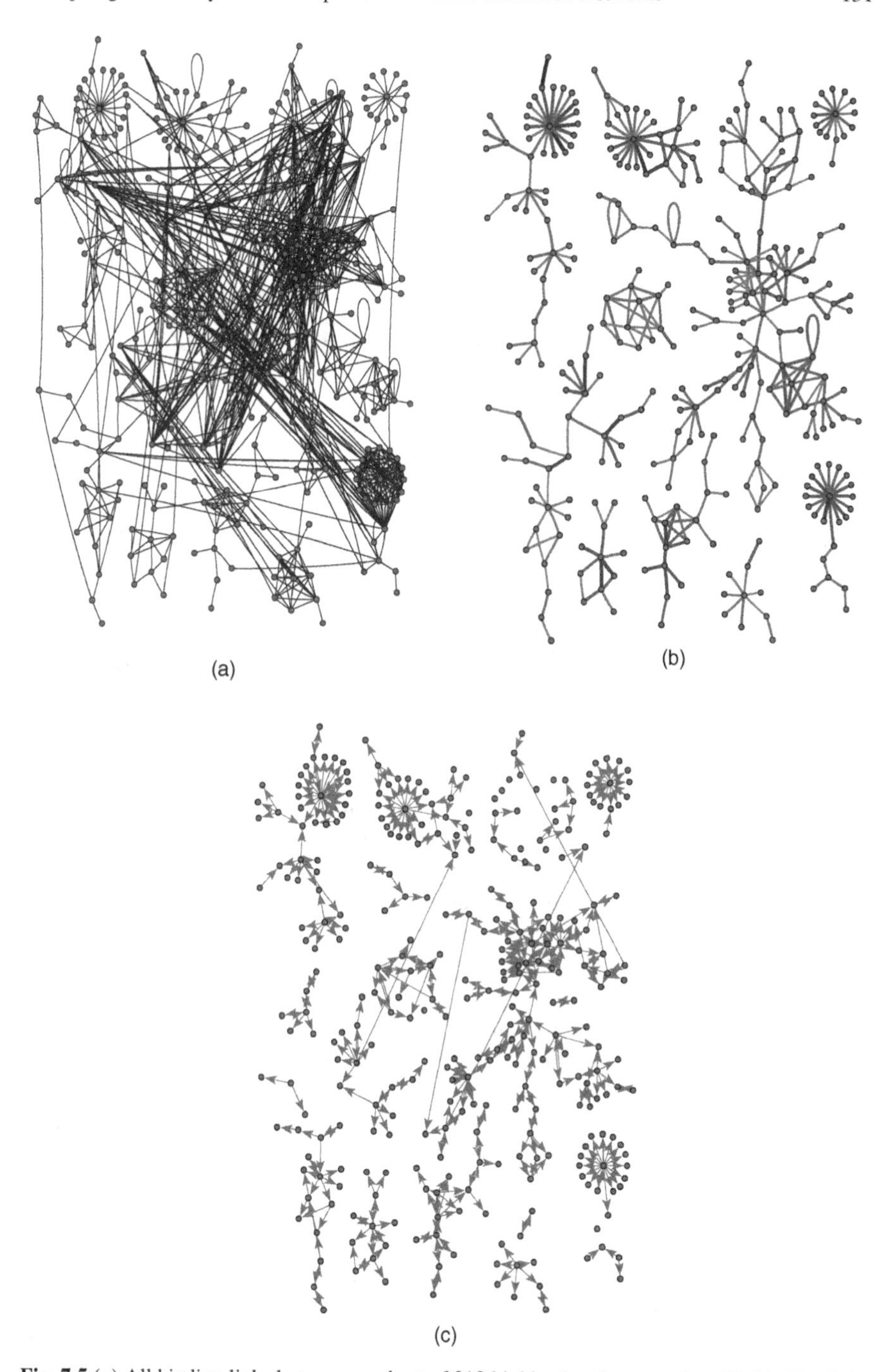

Fig. 7.5 (**a**) All binding links between a subset of 312 highly abundant proteins. (**b**) Binding links characterized by high concentration of heterodimers (> 1000 molecules/cell). The level of gray of binding links scales with the logarithm of concentration of the corresponding heterodimer. (**c**) Concentration-coupled proteins A → B with the property that a twofold increase in the

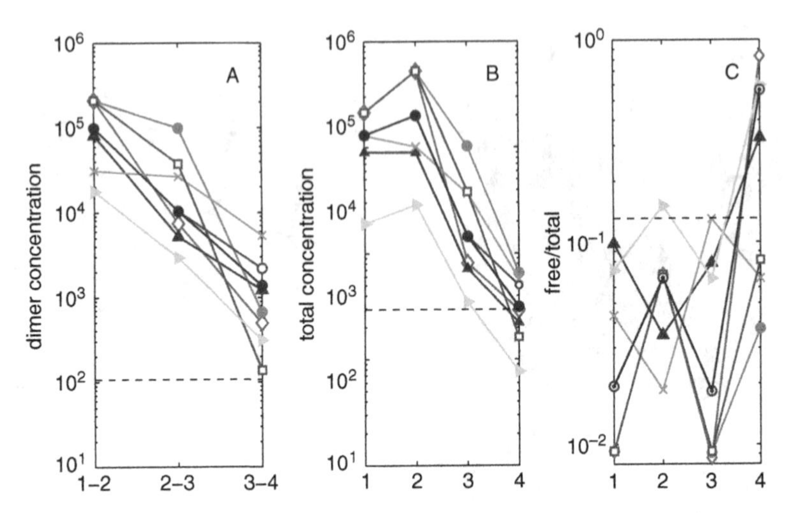

Fig. 7.6 Parameters of the eight three-step chains that exhibit the best transduction of concentration changes as described in the text. The bound (dimer) concentrations D_{ij} (A), total concentrations C_i (B), and free-to-total concentration ratios F_i/C_i (C) of all dimers and proteins involved in these cascading changes are plotted versus the position along the chain (1 being the initially perturbed protein). Different symbols mark eight different highly-transducing chains. Dashed lines correspond to network-wide geometric averages of the corresponding quantities: $\langle D_{ij} \rangle \sim 100$ copies/cell, $\langle C_i \rangle \sim 3000$ copies/cell, and $\langle F_i/C_i \rangle = 13\%$

- Intermediate proteins have to be highly sequestered. That is to say, in order to reduce buffering effects free-to-total concentration ratios F_i/C_i should be sufficiently low for all but the last protein in the chain.
- Total concentrations C_i should gradually decrease in the direction of propagation. Thus propagation of perturbations along virtually all of these long conduits is unidirectional and follows the gradient of concentration changes (a related concept of a "gradient network" was proposed for technological networks in Ref. [42]).
- Free concentrations F_i should alternate between relatively high and relatively low values in such a way that free concentrations of proteins at steps 2 and 4 have enough "room" to go down. The two apparent exceptions to this rule visible in Fig. 7.6 may be optimized to respond to a drop (instead of increase) in the level of the first protein.

These findings are in agreement with our more detailed numerical and analytical analysis of propagation of fluctuations presented in [34]. In [34] we demonstrated that the linear response of the LMA equilibrium to *small* changes in protein abundances could be approximately mapped to a current flow in the resistor network in which heterodimer concentrations play the role of conductivities (which need to be large for a good transmission) while high F_i/C_i ratios result in the net loss of the perturbation "current" on such nodes and thus need to be minimized.

7.3.5 Robustness with Respect to Assignment of Dissociation Constants

It has been often conjectured that the qualitative dynamical properties of biological networks are to a large extent determined by their topology rather than by quantitative parameters of individual interactions such as their kinetic or equilibrium constants (for a classic success story see e.g. [43]). Our results generally support this conjecture, yet go one step further: we observe that the response of reversible protein-protein binding networks to large changes in concentrations strongly depends not only on topology but also on abundances of participating proteins. Indeed, perturbations tend to preferentially propagate via paths in the network in which abundances of intermediate proteins monotonically decrease along the path (see Fig. 7.5). Thus by varying protein abundances while strictly preserving the topology of the underlying network, one can select different conduits for propagation of perturbations.

On the other hand our results indicate that these conduits are to a certain degree insensitive to the choice of dissociation constants. In particular, we found (see Fig. 7.7) that equilibrium concentrations of dimers and the remaining free (unbound) concentrations of individual proteins calculated for two different K_{ij} assignments (K_{ij} = const = 5nM and $K_{ij} = \max(C_i, C_j)/20$ with the inverse mean of 5nM) had a high Spearman rank correlation coefficient of 0.89 and even higher linear Pearson correlation coefficient of 0.98. The agreement was especially impressive in the upper part of the range of dimer concentrations (see Fig. 7.7). For example, the typical difference between dimer concentrations above 1000 molecules/cell was measured to be as low as 40%. As we demonstrated above it is exactly these highly abundant heterodimers that form the backbone for propagation of concentration perturbations. Thus it should come as no surprise that sets of concentration-coupled protein pairs observed for different K_{ij} assignments also have a large ($\sim$ 70–80%) overlap with each other. Such degree of robustness with respect to quantitative parameters of interactions can be partially explained by the following observation: proteins whose abundance is higher than the sum of abundances of all of their binding partners cannot be fully sequestered into heterodimers for any assignment of dissociation constants. As we argued above, such proteins with substantial unbound concentrations considerably dampen the propagation of perturbations, and thus cannot participate in highly conductive chains. Another argument in favor of this apparent robustness is based on extreme heterogeneity of wildtype protein abundances (in the dataset of Ref. [37] they span 5 orders of magnitude). In this case concentrations of heterodimers depend more on relative abundances of two constituent proteins than on the corresponding dissociation constant (within a certain range).

In a separate numerical control experiment we verified that the main results of this study are not particularly sensitive to false positives and false negatives in the network topology inevitably present even in the best curated large-scale data. The percentage of concentration-coupled pairs surviving a random removal or addition of 20% of links in the network generally ranges between 60% and 80%.

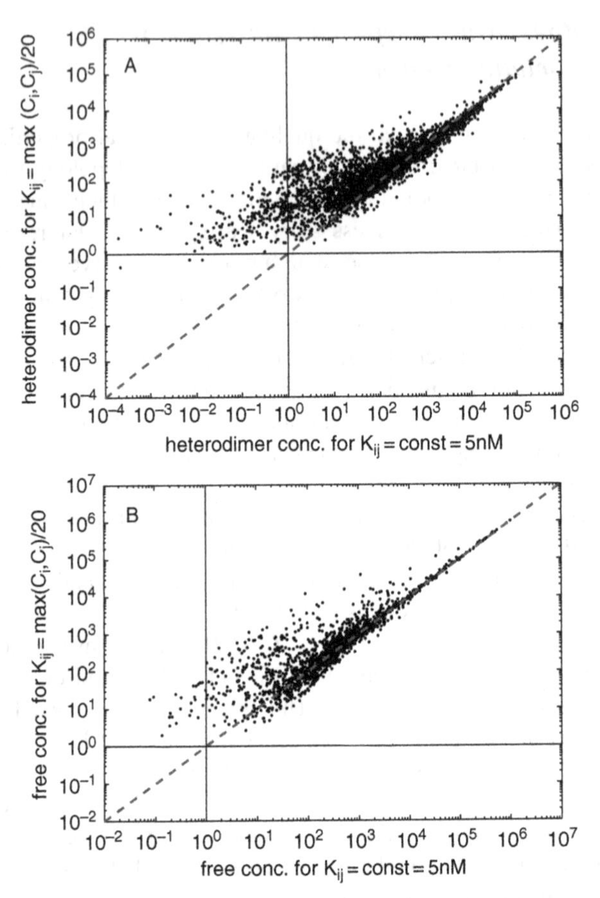

Fig. 7.7 The scatter plot of 4185 bound concentrations D_{ij} (*panel A*) and 1740 values of free concentrations F_i (*panel B*) calculated for two different assignments of dissociation constants to links in the PPI network. The x -axis was computed for the homogeneous assignment $K_{ij} = const = 5\text{nM}$, while the y-axis - for the heterogeneous assignment $K_{ij} = \max(C_i, C_j)/20$ with the same average strength. The *dashed lines* along the diagonals are drawn at $x = y$, while the horizontal and vertical *solid lines* denote the concentration of 1 molecule/cell. Note that equilibrium concentrations in the upper part of their range (e.g. above 1000 molecules/cell) are nearly independent of the choice of K_{ij}. Also, our choice of heterogeneous assignment nearly eliminates free or bound concentrations in a biologically unreasonable range where functional dimers of monomers are present in less than one copy per cell

7.3.6 Effects of Intracellular Noise

Another implication of our findings is for intracellular noise, or small random changes in total concentrations C_i of a large number of proteins. The randomness, smaller magnitude, and the sheer number of the involved proteins characterize the differences between such noise and systematic several-fold changes in the total concentration of one or several proteins considered above. Our methods allow one to

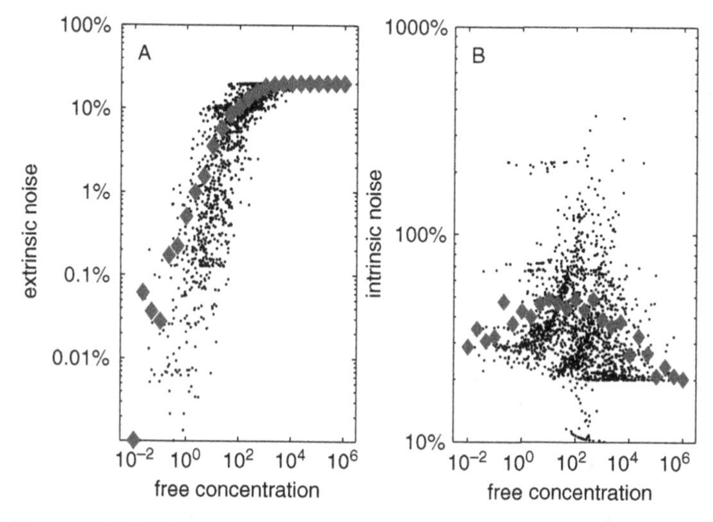

Fig. 7.8 The magnitude of the extrinsic (*panel A*) and the intrinsic (*panel B*) noise in free concentrations F_i of proteins when their total concentrations C_i fluctuate by 20%. *Solid diamonds* show the values of average noise calculated for log-binned free concentrations F_i. In this plot $K_{ij} = const = 1\text{nM}$. One can see that while the extrinsic noise is suppressed in the low-F_i region corresponding to highly sequestered proteins, the intrinsic noise is uniformly high and reaches as much as >300% in the mid-F_i range

decompose the noise in total abundances of proteins into biologically meaningful components (free concentrations and bound concentrations within individual protein complexes). Given a fairly small magnitude of fluctuations in protein abundances (on average around 20% [41]), one could safely employ a computationally-efficient linear response algorithm (see [34]). Several recent studies [44], [45], [41] distinguish between the so-called extrinsic and intrinsic noise. The extrinsic noise corresponds to synchronous or correlated shifts in abundance of multiple proteins that, among other things, could be attributed to variation in cell sizes and their overall mRNA and protein production or degradation rates. Conversely, the intrinsic noise is due to stochastic fluctuations in production and degradation and thus lacks correlation between different proteins. We found that extrinsic and intrinsic noise affect equilibrium concentrations of proteins in profoundly different ways (see Fig. 7.8). In particular, while multiple sources of the extrinsic noise partially (yet not completely) cancel each other, intrinsic noise contributions from several sources can sometimes add up and cause considerable fluctuations in equilibrium free and bound concentrations of particular proteins.

Acknowledgments Work at Brookhaven National Laboratory was carried out under Contract No. DE-AC02-98CH10886, Division of Material Science, U.S. Department of Energy. This work was supported in part by 1 R01 GM068954-01 grant from NIGMS and National Science Foundation under Grant No. PHY05-51164.

References

1. Uetz P, Giot L, Cagney G, Mansfield TA, Judson RS, et al. (2000) A comprehensive analysis of protein-protein interactions in Saccharomyces cerevisiae. Nature 403:623–7.
2. Ito T, Chiba T, Ozawa R, Yoshida M, Hattori M, Sakaki Y (2001) A comprehensive two-hybrid analysis to explore the yeast protein interactome. Proc Natl Acad Sci U S A 98:4569–74
3. Rain JC, Selig L, De Reuse H, Battaglia V, Reverdy C, et al. (2001) The protein-protein interaction map of Helicobacter pylori. Nature 409:211–5.
4. Giot L, Bader JS, Brouwer C, Chaudhuri A, Kuang B, et al. (2003) A protein interaction map of Drosophila melanogaster. Science 302:1727–36.
5. Li S, Armstrong CM, Bertin N, Ge H, Milstein S, et al. (2004) A map of the interactome network of the metazoan C. elegans. Science 303:540–3.
6. LaCount DJ, Vignali M, Chettier R, Phansalkar A, Bell R, et al. (2005) A protein interaction network of the malaria parasite Plasmodium falciparum. Nature 438:103–7.
7. Rual JF, Venkatesan K, Hao T, Hirozane-Kishikawa T, Dricot A, et al. (2005) Towards a proteome-scale map of the human protein-protein interaction network. Nature 437:1173–8.
8. Stelzl U, Worm U, Lalowski M, Haenig C, Brembeck FH, et al. (2005) A human protein-protein interaction network: a resource for annotating the proteome. Cell 122:957–68.
9. Gavin AC, Bosche M, Krause R, Grandi P, Marzioch M, et al. (2002) Functional organization of the yeast proteome by systematic analysis of protein complexes. Nature 415:141–7.
10. Ho Y, Gruhler A, Heilbut A, Bader GD, Moore L, et al. (2002) Systematic identification of protein complexes in Saccharomyces cerevisiae by mass spectrometry. Nature 415:180–3.
11. Rual JF, Venkatesan K, Hao T, Hirozane-Kishikawa T, Dricot A, et al. (2005) Towards a proteome-scale map of the human protein-protein interaction network. Nature 437:1173–8.
12. Krogan NJ, Cagney G, Yu H, Zhong G, Guo X, et al. (2006) Global landscape of protein complexes in the yeast Saccharomyces cerevisiae. Nature 440:637–43.
13. Gavin AC, Aloy P, Grandi P, Krause R, Boesche M, et al. (2006) Proteome survey reveals modularity of the yeast cell machinery. Nature 440:631–6.
14. Barabasi AL and Albert R (1999) Emergence of scaling in random networks. Science 286:509–12.
15. Jeong H, Mason SP, Barabasi AL and Oltvai ZN (2001) Lethality and centrality in protein networks. Nature 411:41–2.
16. Maslov S and Sneppen K (2002) Specificity and stability in topology of protein networks. Science 296:910–3.
17. Wagner A (2001) The yeast protein interaction network evolves rapidly and contains few redundant duplicate genes. Mol Biol Evol 18:1283–92.
18. Spirin V and Mirny LA (2003) Protein complexes and functional modules in molecular networks. Proc Natl Acad Sci U S A 100:12123–8.
19. Shi YY, Miller GA, Qian H, and Bomsztyk K (2006) Free-energy distribution of binary protein-protein binding suggests cross-species interactome differences. Proc Nat Acad of Sci U S A 103:11527–32.
20. Evlampiev K and Isambert H 2006. Asymptotic Evolution of Protein-protein Interaction Networks for General Duplication-Divergence Models. Preprint q-bio.MN/0611070 at arxiv.org.
21. Vazquez A, Flammini A, Maritan A, and Vespignani A (2001) Modelling of protein interaction networks. Preprint cond-mat/0108043 at arxiv.org. Published in (2003) ComPlexUs 1:38.
22. Sole R V, Pastor-Satorras R, Smith E, and Kepler TB (2002) A model of large-scale proteome evolution, Preprint cond-mat/0207311 at arxiv.org. Published in (2002) Advances in Complex Systems 5:43.
23. Ispolatov I, Krapivsky PL, and Yuryev A (2005) Duplication-divergence model of protein interaction network. Phys Rev E 71:061911.
24. Caldarelli G, Capocci A, De Los Rios P, and Munoz MA (2002) Scale-free networks from varying vertex intrinsic fitness. Phys Rev Lett 89:258702.
25. Deeds EJ, Ashenberg O, and Shakhnovich EI (2006) A simple physical model for scaling in protein-protein interaction networks. Proc Nat Acad Sci U S A 103(2):311–6.

26. Maslov S and Ispolatov I (2007) Propagation of large concentration changes in reversible protein-binding networks. Proc Natl Acad Sci U S A 104:13655–60.
27. Kannan R, Tetali P, and Vempala S. (1999) Simple Markov-chain algorithms for generating bipartite graphs and tournaments. Random Structures and Algorithms 14:293–308.
28. The set of MATLAB programs can be downloaded at http://www.cmth.bnl.gov/maslov/matlab.htm
29. Maslov S, Sneppen K, and Zaliznyak A (2002) Pattern Detection in Complex Networks: Correlation Profile of the Internet. Preprint cond-mat/0205379 at arxiv.org; published in Physica A 333:529–540.
30. Watts D and Strogatz, S (1998) Collective dynamics of small world networks. Nature 293: 400–403.
31. Shen-Orr S, Milo R, Mangan S, and Alon U (2002) Network motifs in the transcriptional regulation of Escherichia coli. Nature Genetics, 31:64–68.
32. Milo R, Shen-Orr S, Itzkovitz S, et al. (2002) Network motifs: simple building blocks of complex networks. Science 298:824–7.
33. Maslov S and Sneppen K (2002) Protein interaction networks beyond artifacts. FEBS Letters 530:255–6.
34. Maslov S, Sneppen K, Ispolatov I (2007) Spreading out of perturbations in reversible reaction networks. New Journal of Physics 9:273(11 pages).
35. Albert R, Jeong H, and Barabasi AL (2000) Error and attack tolerance of complex networks. Nature 406:378–82.
36. Stark C, Breitkreutz BJ, Reguly T, Boucher L, Breitkreutz A, and Tyers M (2006) BioGRID: a general repository for interaction datasets. Nucleic Acids Res 34:D535–9.
37. Ghaemmaghami S, Huh WK, Bower K, Howson RW, Belle A, Dephoure N, O'Shea EK, and Weissman, JS (2003) Global analysis of protein expression in yeast. Nature 425:737–41.
38. Piehler J (2005) New methodologies for measuring protein interactions in vivo and in vitro. Curr Opin in Struct Biol 15:4–14.
39. Kumar MD and Gromiha MM (2006) PINT: Protein-protein Interactions Thermodynamic Database. Nucleic Acids Res 34:D195–8.
40. Lancet D, Sadovsky E, and Seidemann E (1993) Probability model for molecular recognition in biological receptor repertoires: significance to the olfactory system. Proc Natl Acad Sci U S A 90(8):3715–9.
41. Newman JRS, Ghaemmaghami S, Ihmels J, Breslow DK, Noble M, DeRisi JL, and Weissman JS (2006) Single-cell proteomic analysis of S. cerevisiae reveals the architecture of biological noise. Nature 441:840–6.
42. Toroczkai Z and Bassler KE (2004) Jamming is limited in scale-free systems. Nature 428:170.
43. vonDassow G, Meir E, Munro EM, and Odell GM (2000) The segment polarity network is a robust developmental module. Nature 406:188–92.
44. Elowitz MB, Levine AJ, Siggia ED, and Swain PS (2002) Stochastic gene expression in a single cell. Science 297:1183.
45. Raser JM and O'Shea EK (2005) Noise in gene expression: origins, consequences, and control. Science 309:2010–3.

Chapter 8
From Protein Interaction Networks to Protein Function

Mona Singh

Abstract The recent availability of large-scale protein-protein interaction data provides new opportunities for characterizing a protein's function within the context of its cellular interactions, pathways and networks. In this paper, we review computational approaches that have been developed for analyzing protein interaction networks in order to predict protein function.

8.1 Introduction

A major challenge in the post-genomic era is to determine protein function at the proteomic scale. Most organisms contain a large number of proteins whose functions are currently unknown. For example, about one-third of the proteins in the baker's yeast *Saccharomyces cerevisiae*—arguably one of the most well-characterized model organisms—remain uncharacterized. Traditionally, computational methods to assign protein function have relied largely on sequence homology. However, the recent emergence of high-throughput techniques for determining protein interactions has enabled a new line of research where protein function is predicted by utilizing interaction data.

Proteome-scale physical interaction networks, or interactomes, have been determined for several organisms, including yeast and human. These networks are comprised of direct physical interactions between proteins (typically obtained via two hybrid analysis [FS89]) as well as of interactions indicating that two proteins are part of the same multi-protein complex (review, [BK03]). High-throughput experiments have also linked together proteins in several other ways, and it is possible to build large-scale networks consisting of links between proteins that are synthetic lethals or are coexpressed, or between proteins where one regulates or phosphorylates the

M. Singh
Department of Computer Science and Lewis-Sigler Institute for Integrative Genomics, Princeton University 08544, USA
e-mail: mona@cs.princeton.edu

A. Panchenko, T. Przytycka (eds.), *Protein-protein Interactions and Networks*,
DOI: 10.1007/978-1-84800-125-1_8, © Springer-Verlag London Limited 2008

other (review, [ZGS07]). In addition to interaction networks that have been determined experimentally, there are a number of computational methods for building functional interaction networks, where two proteins are linked if they are predicted to perform a shared biological task (review, [GK00])).

In this chapter, we review some of the basic computational methods developed for analyzing protein interaction networks in order to predict protein function. The majority of these methods use some version of *guilt-by-association*, where proteins are annotated by transferring the functions of the proteins with which they interact. The methods differ in the extent to which they use global properties of the interactome in annotating proteins, what topological features of the interactome they exploit, and whether they rely on first identifying tight clusters of proteins within the interactome before transferring annotations. Additionally, the underlying formulations are quite diverse, typically exploiting and further developing well understood concepts from graph theory, graphical models, discriminative learning and/or clustering.

While there are many sorts of protein interaction networks, we will largely limit our discussion to networks comprised of physical interactions between proteins. It is often straightforward to apply the methods overviewed in this chapter to other types of interaction networks; however, performance of methods is expected to vary on different types of networks, perhaps dramatically, as the underlying topological features of these networks can be different. We refer the reader to other excellent reviews [AS06, SUS07] for alternate viewpoints that additionally consider function prediction methods that integrate physical interaction networks with data from other experimental sources.

8.2 Preliminaries

8.2.1 Protein Function

Protein function is an abstract notion that can mean different things. The Gene Ontology (GO) [ABB00] classifies function into separate categories, each of which contains a directory of terms and specifies the relationships between them. The two categories referring most directly to function are molecular function and biological process. The molecular function of a protein describes its biochemical activity, whereas its biological process specifies the role it plays in the cell or the pathway in which it participates. Additionally, GO organizes terms relating to location; the cellular component category has terms which refer to the places where the protein is found. These views of protein function are largely orthogonal: for example, proteins with the same molecular function can play a role in different pathways, and a pathway is built of proteins of various molecular functions. From the perspective of function prediction, molecular functions, which correspond to the intrinsic features of the protein, are often predicted based on sequence or structural similarity to proteins of known function, whereas biological processes, being fundamentally collaborative, are often predicted based on a protein's functional interaction partners (e.g., the proteins with which it is co-expressed or the proteins with which it interacts physically). In this chapter, when we refer to a protein's function, we will typically mean its biological process, though a protein's cellular component may also be effectively predicted based on its interaction partners via guilt-by-association.

8.2.2 Notation

As elsewhere in this book, a protein-protein interaction network is represented as a graph $G = (V, E)$, where there is a vertex $v \in V$ for each protein, and an edge (u, v) between two vertices u and v if the corresponding proteins interact. Throughout the chapter, self-interactions are ignored. Let N denote the number of proteins in the network. The network can also be represented by its $N \times N$ adjacency matrix A, where $A_{uv} = 1$ if $(u, v) \in E$ and 0 otherwise. Let $\mathcal{F}$ be the set of possible protein functional annotations. Each protein may be annotated with one or more annotations from $\mathcal{F}$. That is, each vertex $v \in V$ may have a set of labels associated with it. The edges in the network may be weighted; typically the weight $w_{u,v}$ on the edge between u and v reflects how confident we are of the interaction between u and v. If each interaction given in the network is considered equally trustworthy, the network may be considered unweighted or with unit-weighted edges.

Many approaches discussed below utilize the "neighborhood" of a protein. Let $\mathcal{N}_r(u)$ denote the neighborhood of protein u within radius r; that is, $\mathcal{N}_r(u)$ is the set of proteins where each protein has some path in the network to u that is made up of at most r edges. Thus, $\mathcal{N}_0(u)$ consists of protein u, $\mathcal{N}_1(u)$ consists of protein u and all proteins that interact with u, $\mathcal{N}_2(u)$ consists of the proteins in $\mathcal{N}_1(u)$ along with all proteins that interact with any of the proteins in $\mathcal{N}_1(u)$, and so on.

8.3 Assessing Interaction Reliability

Before delving into the various methods that have been developed for analyzing interactomes in order to predict protein function, we briefly discuss the important issue of network reliability. In particular, it is well known that high-throughput physical interaction data are noisy, and that reliability of different data sources vary, even if they are based on the same underlying technology (e.g., see [vMKS+02, DSC03, SSM03]). Weighted networks are thus useful in the context of protein function prediction, as weights can be chosen to model the reliability of each physical interaction. Here, we review a simple scheme for assessing physical interaction reliability [NJA+05], that is essentially the same as the ones used in several approaches for data integration [vMHJ+03, JCB+04].

For each experimental source i (e.g., each high-throughput data set may be considered as coming from a single source), let r_i denote the probability that an interaction observed in this experiment is a true physical interaction. Assuming that the observations from each experimental source are independent, it is possible to estimate the probability of a physical interaction between proteins u and v as:

$$1 - \Pi_i(1 - r_i),$$

where the product is taken over all experiments i where an interaction between u and v is found. This estimate can be used as the weight $w_{u,v}$ of the edge between u and v. If r_i is chosen to be identical for all experimental sources, this approach simply gives higher reliability to physical interactions that have been observed multiple times. Alternatively, for each experimental source i, r_i can be estimated by computing, for example, the fraction of its interactions that connect proteins with a known shared function. It has been shown that a wide range of network analysis algorithms perform better in predicting protein function when utilizing this scheme for assessing interaction reliability than when considering all interactions as equally likely [NJA+05, CSW06].

There are also several probabilistic data integration schemes that combine many different types of data (e.g., expression, localization and physical interaction) in order to functionally link proteins (e.g., [vMHJ+03, TDO+03, JYG+03, LDAM04]). Each link is associated with a weight that represents the probability, or some other confidence measure, that the two corresponding proteins are functionally related. The resulting weighted networks can then be used for protein function prediction. Though closely related, note that predicting functional linkages is not the same as predicting the function of a protein, as each protein can be linked with varying levels of confidence to several proteins with multiple biological process annotations; some method or rule is still necessary to decide which annotations are transferred.

While in this chapter we often describe the basic algorithms in the context of unweighted networks, many of the methods discussed have been extended in a straightforward manner to incorporate weighted edges.

8.4 Algorithms

Numerous computational methods have been developed for functionally annotating proteins using interaction networks. Our discussion below categorizes methods based upon their underlying formulations and algorithmic solutions.

8.4.1 Local Approaches

The simplest method for predicting the function of a protein from an interaction map is to assign to each protein the biological process that is most frequent among its direct interactions [SUF00]. In this case, the score for annotating a protein u with a particular annotation a could be the number (or alternatively the fraction) of proteins that u interacts with that are annotated with a. In the case of weighted interaction networks, a weighted sum can be used instead. This majority or neighborhood-counting approach is purely local, and takes only limited advantage of the underlying graph structure of the network. Subsequent graph-theoretic approaches have attempted to generalize this principle to consider linkages beyond the immediate neighbors in the interaction graph, both to provide a systematic framework for analyzing the entirety of physical interaction data for a given proteome, and to make predictions for proteins with no annotated interaction partners.

One way to extend the majority approach is by predicting a protein's function by looking at all proteins within a neighborhood of specified radius and finding over-represented functional annotations [HNO+01]. For each protein u and a fixed radius r, this neighborhood approach considers all proteins in $\mathcal{N}_r(u)$ (i.e., within radius r of u) and then for each function, computes a score based on the χ^2 test. Specifically, the score is computed as $\frac{(f-e)^2}{e}$, where f is the number of proteins with the function under consideration within the neighborhood and e is the number of proteins expected to have that function within the neighborhood, given the frequency of the function in the entire interactome. The function with the highest χ^2 score is assigned to the protein. With radius one, this approach is similar to the simpler majority approach; however, if two functions annotate the same number of a protein's direct neighbors, the neighborhood approach would favor the one that annotates fewer proteins in the entire interactome. While this approach moves beyond direct neighbors, it does not consider any aspect of network topology within the local neighborhood. For example, Fig. 8.1 shows two interaction networks that are treated equivalently when considering a radius of two and annotating protein u; however, in the first case, all paths that connect protein u to the annotated proteins share a common link (between u and v), and in the second case, there are several independent paths between u and the annotated proteins, and moreover, two of these proteins are directly adjacent to u. Perhaps because this method completely ignores network topology within neighborhoods, its biological process predictions are best when considering neighborhoods of radius one [HNO+01]. A recent extension attempts to include proteins at radius two by assigning weights to each protein in the

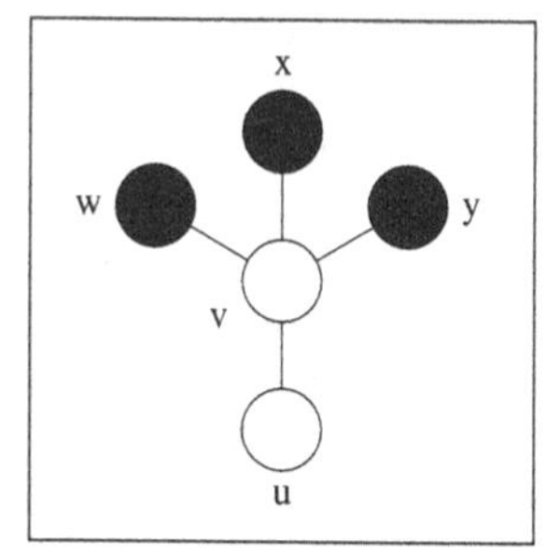

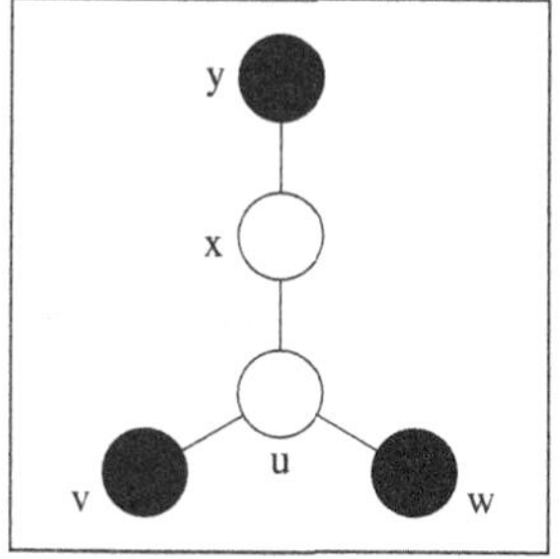

Fig. 8.1 Two protein interaction graphs, where *dark colored nodes* correspond to proteins that are known to take part in the same process, and *light colored nodes* do not have biological process annotations. When annotating protein u, a neighborhood approach [HNO+01] with radius two would treat these networks equivalently, despite the stronger evidence in the second network for protein u to be annotated with the function of the proteins corresponding to the *dark colored nodes*

neighborhood by favoring the number of shared interactions it has with the protein being annotated, and then scoring each function based on its weighted frequency in the neighborhood [CSW06]; intuitively, this weighting scheme has similarities to the Czekanowski-Dice distance described in Section 4.6.

8.4.2 Graph Cuts

Since in many cases the local neighborhood of a protein may not contain enough annotated proteins to determine protein function well, global methods that consider the entire network and its annotations are necessary. Several methods have been proposed that exploit the global topological structure of the interaction network by annotating proteins so as to minimize the number of times different annotations are associated with interacting proteins [VFMV03, KML+04, NJA+05]. The functions can be considered simultaneously [VFMV03, NJA+05], or just one at a time [KML+04].

If all functions are considered at the same time, this optimization problem is a generalization of the computationally difficult minimum multiway k-cut problem [DJP+92], where the goal is to partition a graph in such a way that each of k terminal nodes belongs to a different subset of the partition and so that the (weighted) number of edges that are "cut" in the process is minimized. In the more general version of the multiway k-cut problem relevant to protein functional annotation, the goal is to assign a function to all the unannotated nodes so as to minimize the sum of the weights of the edges joining nodes with no function in common. In the case where one function is considered at a time, each protein that is known to have that function is labeled as a "positive" and each protein that is known to have some function but not the one being considered is labeled as a "negative." It is straightforward to see that this formulation of the problem can be stated as a minimum cut problem, and thus exact solutions are obtainable in polynomial time (e.g., see [CLR90]).

Several techniques have been applied to solve these cut problems for interactomes. In the case where one function at a time is considered, a deterministic approximation algorithm has been applied to obtain a single solution per function [KML+04]. In this application, a version is also considered where edges are assigned (positive) weights based on the correlation of the corresponding proteins' expression profiles. In subsequent work, this formulation has been solved exactly using a minimum cut algorithm [MWK+06]. In the case where multiple functions are considered at once, simulated annealing has been applied and solutions from several runs have been aggregated [VFMV+03]. That is, the score of a function for a particular protein is given by the number of runs in which the simulated annealing solution annotates the protein with the function. The simulated annealing approach is a heuristic and thus does not guarantee an optimal solution to the underlying optimization problem.

An integer linear programming (ILP) formulation for the generalized multiway cut problem has also been proposed [NJA+05]. While ILP is computationally difficult from a theoretical point of view, in practice optimal solutions to this ILP, and thus the original optimization problem, can be readily obtained for existing physical interactomes using AMPL [FGK02] and the commercial solver CPLEX [ILO00]. In this formulation, there is a node variable $x_{u,a}$ for each protein $u \in V$ and each possible functional annotation a in the set of functions $\mathcal{F}$. This variable will be set to 1 if protein u is predicted to have function a. If a protein u has known functional annotations, variables $x_{u,a}$ are fixed as 1 for its known annotations a and as 0 for all other annotations. There is also an edge variable $x_{u,v,a}$ for each possible functional annotation a and each pair of interacting proteins u and v. This variable is set to 1 if both proteins u and v are annotated with function a. Minimizing the weighted number of neighboring proteins with different annotations is the same as maximizing the number with the same annotation, and so we have the following ILP:

$$
\begin{array}{ll}
\text{maximize} & \sum_{(u,v)\in E, a\in\mathcal{F}} x_{u,v,a} w_{u,v} \\
\text{subject to} & \\
\sum_a x_{u,a} = 1 & \text{if } annot(u) = \emptyset \\
x_{u,a} = 1 & \text{if } a \in annot(u) \\
x_{u,a} = 0 & \text{if } a \notin annot(u), annot(u) \neq \emptyset \\
x_{u,v,a} \leq x_{u,a} & \text{for } (u, v) \in E \text{ and } a \in \mathcal{F} \\
x_{u,v,a} \leq x_{v,a} & \text{for } (u, v) \in E \text{ and } a \in \mathcal{F} \\
x_{u,v,a}, x_{u,a} \in \{0, 1\} & \text{for all } u, v \text{ and } a.
\end{array}
\tag{8.1}
$$

Here, $annot(u)$ is the set of known annotations for protein u. The final constraint is the integrality constraint. The first constraint specifies that exactly one functional annotation is made for any protein. The second and third constraints ensure that if protein u is annotated with function a, $x_{u,a}$ is set as a constant to 1, and if protein u is annotated but not with function a, $x_{u,a}$ is set as a constant to 0. The fourth and fifth constraints ensure that a particular function is picked for an edge only if it is also chosen for the corresponding proteins.

An important consideration in this framework is the existence of multiple optimal solutions. For example, the network in Fig. 8.2 has seven minimum cuts of value one, and the cut criterion does not favor any one cut over the other. However, we expect proteins that are closer together in the network to have more similar biological process annotations than those that are further apart. Thus, in the network in Fig. 8.2, we would want the proteins closer to v_1 to be annotated with its function a_1, and the proteins closer to v_8 to be annotated with its function a_2. If we find all optimal cuts for this graph, we observe that v_2 is in fact annotated with a_1

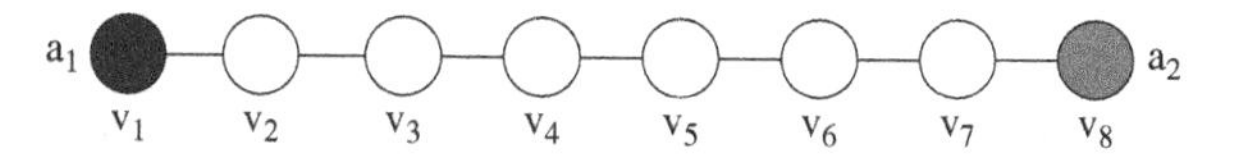

Fig. 8.2 Proteins v_1 and v_8 are annotated with functions a_1 and a_2, respectively. There are seven ways to annotate proteins so that there is only one edge that connects proteins with different annotations. However, proteins further away from protein v_1 are less likely to have function a_1 than those closer to v_1. A single cut of the graph does not take into account such distance effects

more often than with a_2 in the assignments made by these cuts. Thus, a sense of distance to annotated nodes can be present in the set of all optimal solutions. In the simulated annealing approach to this problem, information from multiple solutions is utilized [VFMV03]. If each run does indeed converge to an optimal solution, considering multiple runs amounts to sampling from the space of optimal solutions. The ILP can also be modified to find multiple solutions [NJA+05]. One approach is based on solution exclusion, and constraints are added to the ILP that require each consecutive solution to be different from any previous one in the values it assigns to at least 5% of the node variables. Alternatively, it is possible to change the formula to include uniform self-weights $w_{u,a}$ for each protein u and function a. These self-weights are then perturbed by adding a very small random offset to each. The objective function is changed to:

$$\text{maximize} \sum_{(u,v)\in E, a\in\mathcal{F}} x_{u,v,a} w_{u,v} + \sum_{u\in V, a\in\mathcal{F}} x_{u,a} w_{u,a},$$

subject to the same set of constraints. The perturbation in weights is too small to change the solution to the underlying problem, but it does cause the solver to choose a different optimal solution. The score for a function for a protein is then the number of times this function is assigned to the protein among the obtained solutions.

8.4.3 Markov Random Field

Markov random field approaches have also been introduced for the problem of predicting protein function from interaction networks [DZM+03, DCS03, LK03]. A Markov random field is an undirected graphical model that represents the joint probability distribution of a set of random variables. It is specified by an undirected graph where each vertex represents a random variable and each edge represents a dependency between two random variables, such that the state of any random variable is independent of all others given the states of its neighbors. The joint distribution represented by a Markov random field is computed by considering a potential function over each of its cliques. That is, for N random variables X_i, the probability of an assignment of the states is given by:

$$\Pr(X_1 = x_1, \ldots, X_N = x_N) = \frac{1}{Z} e^{-\sum_k \Phi_k(X_{\{k\}})},$$

where k enumerates all cliques, Φ_k is the potential function associated with the k-th clique, $X_{\{k\}}$ gives the states of the k-th clique's random variables, and Z is a normalizing constant.

In an application to protein function annotation via network analysis [DZM+03, DCS03], one function at a time is considered. Each protein has a random variable associated with it, and its state corresponds to whether the function under consideration is assigned or not. It is assumed that the joint distribution can be expressed as an expression that considers only cliques of size at most two (i.e., edges). This means that the potential function evaluating the network is a linear expression composed of terms over the vertices and edges. So,

$$\Pr(X_1 = x_1, \ldots, X_N = x_N) = \frac{1}{Z} e^{-(\sum_{u \in V} \phi_1(X_{\{u\}}) + \sum_{(u,v) \in E} \phi_2(X_{\{u,v\}}))},$$

where ϕ_1 computes the vertex "self-term" and the ϕ_2 computes the pairwise edge term. The self-term potential is chosen to correspond to the prior probability for annotating a protein with a particular function; it takes into account the frequency of the function in the network. The pairwise edge potential is chosen to have different values corresponding to the three cases where either the interacting proteins both have the function under consideration, or they both do not have that function, or one has that function and the other does not; these values are determined using a quasi-likelihood method. Note that these values are not necessarily the same for each function. The posterior probability that a protein has the function of interest is computed using Gibbs sampling, and then if this value is above a chosen threshold, the function is predicted for the protein. As noted earlier [DTSC04], this model is a generalization of the per-function cut-based method [KML$^+$04], and is similar to that of the multiple function cut formulation [VFMV03]. In particular, the cut-based models assume the same fixed value for interactions between proteins of the same function (or for interactions between a protein of one function and any other), regardless of function; this may not be the best assumption, as different functions may transfer annotations to varying degrees.

A somewhat different Markov random field approach for protein function annotation has also been proposed [LK03]. In particular, it is assumed that the number of neighbors of a protein that have a particular functional annotation is binomially distributed according to a parameter that differs depending on whether the protein has that function or not. The posterior probabilities for each protein are computed via a heuristic version of belief propagation (review, [YFW03]).

8.4.4 Network Flow-Based Methods

While cut-based approaches take into account more global properties of interaction maps, they do not reward local proximity in the network. For example, if only two proteins have annotations in a particular network, all other proteins will be labeled by one of these annotations, regardless of the size of the network. One attempt to overcome this problem uses the idea of network flow [NJA$^+$05]. Network flow is dual to the notion of graph cut (e.g., see [CLR90]); however, here network flow is used as the inspiration for a simulation method. Informally, each protein of known functional annotation is a "source" of "functional flow" that can be propagated to unannotated nodes, using the edges in the interaction graph as a conduit. Each protein has a "reservoir" which represents the amount of flow that the node can pass on to its neighbors at the next iteration, and each edge has a capacity (its weight) limiting the amount of flow that can pass through the edge in one iteration. Each iteration of the algorithm updates the reservoirs using simple local rules described below. The simulation is run for a fixed number of steps, and a functional score for each protein is obtained by summing the total amount of flow for that function that the protein has received over the course of the simulation. This method incorporates a notion of distance in the network as the effect of each annotated protein on any other protein decreases with increasing distance between them. Network connectivity is exploited, as multiple disjoint paths between functional sources and a protein results in more flow to the protein.

More formally, for each protein u in the interaction network, there is a variable $R_t^a(u)$ that corresponds to the amount in the reservoir for function a that node u has at time t. For each edge (u, v) in the interaction network, there are variables $g_t^a(u, v)$ and $g_t^a(v, u)$ that represent the flow of function a at time t from protein u to protein v, and from protein v to protein u. The algorithm runs for d time steps or iterations. At time zero, there are only reservoirs of function a at annotated nodes:

$$R_0^a(u) = \begin{cases} \infty & \text{if } u \text{ is annotated with } a \\ 0 & \text{otherwise.} \end{cases}$$

Since reservoirs are infinite in the source proteins, the sources always have enough flow in their reservoir to fill the capacity of their outgoing edges. At each subsequent time step, the reservoir of each protein is computed by considering the amount of flow that has entered the node and the amount that has left:

$$R_t^a(u) = R_{t-1}^a(u) + \sum_{v:(u,v)\in E} (g_t^a(v, u) - g_t^a(u, v)).$$

Initially, at time 0, there is no flow on the edges, and $g_0^a(u, v) = 0$. At each subsequent time step, flow proceeds “downhill” and satisfies the capacity constraints. That is, flow only spreads from proteins with more filled reservoirs to nodes with less filled reservoirs. Moreover, a node pushes the flow residing in its reservoir to its neighbors proportionally to the capacities of the respective edges. Specifically, the flow over edges at time $t > 0$ is given by:

$$g_t^a(u, v) = \begin{cases} 0, \text{ if } R_{t-1}^a(u) < R_{t-1}^a(v) \\ min\left(w_{u,v}, R_{t-1}^a(u)\frac{w_{u,v}}{\sum_{(u,y)\in E} w_{u,y}}\right), & \text{otherwise.} \end{cases}$$

Finally, the functional score for node u and function a over d iterations is calculated as the total amount of flow that has entered the node:

$$f_a(u) = \sum_{t=1}^{d} \sum_{v:(u,v)\in E} g_t^a(v, u).$$

For each protein, the function with the highest score is taken as its prediction, provided this score is greater than some threshold.

Note that a source's immediate neighbor in the graph receives d iterations worth of flow from the source, while a node that is two links away from the source receives $d - 1$ iterations worth of flow. Similarly, the number of iterations for which the algorithm is run determines the maximum number of interactions that can separate a recipient node from a source in order for the flow to propagate from the source to the recipient. For the protein interaction context, a relatively small number of iterations has worked well in practice (e.g., less than half the diameter of the network).

Recently, a similar deterministic flow-based simulation approach has been developed for finding clusters in protein interaction networks [CHRZ07].

8.4.5 Discriminative Learning Methods

An alternate approach for predicting protein function from protein interaction networks is to use machine learning techniques within a classification framework. Support vector machines (SVMs) are machine learning methods that embed positive and negative examples in a feature space and then find a maximal separating hyperplane in this space [Vap98, Bur98]. They have been applied in the context of function prediction via network analysis [LBC+04, TN04], where each function is considered in turn, and each protein is labeled as positive or negative based upon whether it is annotated with the function of interest. The key technical issue is how each protein u in the network is mapped into a point x_u in the feature space. If proteins are "close" in the network, then this mapping should ensure that they are also close in the feature space. The mapping can be given implicitly via a positive definite kernel matrix K. This kernel matrix specifies the inner product (i.e., $K_{uv} = x_u^T x_v$); since the discriminant function for SVMs is specified via inner products, explicit representations of the points are not necessary.

In [LBC+04], two kernels are considered. First, a linear kernel is created where each entry K_{uv} corresponds to the dot product of the N-dimensional vectors representing the interactions of proteins u and v. The more similar the interaction patterns for the proteins, the larger this value is in the kernel matrix; this kernel does not capture more global properties of the network. Second, a diffusion kernel [KL02] is created where the kernel value K_{uv} can be interpreted as the probability that a random walk starting from u will be at v after infinite time steps; the transition probabilities between nodes are dependent on a parameter specifying the rate of diffusion. The diffusion kernel accounts for all possible paths connecting two proteins, with nodes that are connected with shorter paths or by several paths considered more similar. It has been shown that the diffusion kernel captures the global constraint that the sum of the Euclidean distances between connected samples is bounded, but that this can lead to large variances in the pairwise distances [TN04]. This observation has led to the development of a locally constrained diffusion kernel, which captures additional local constraints requiring that the Euclidean distance between connected samples be more tightly bound. SVMs using the locally constrained diffusion kernel are found to better predict protein function than those using the original diffusion kernel.

8.4.6 Clustering

The approaches reviewed above exploit the idea that neighboring proteins tend to have similar biological processes. More broadly, cellular networks have been proposed to be organized in a modular fashion [HHLM99]. These modules correspond to sets of proteins that take part in the same cellular function or together comprise a protein complex. One general class of approaches for predicting protein function thus attempts to cluster protein interaction networks in order to uncover these functional modules. These functional modules, or clusters, are useful for annotating uncharacterized proteins, as the most common functional annotation within a cluster can be transferred to its uncharacterized proteins. An alternate approach for transferring annotations uses the hypergeometric distribution to determine, for each function, whether it is enriched in a cluster, and if so, to transfer this function to all proteins in the cluster. Such an approach computes a p-value for a particular function in a cluster as:

$$p = 1 - \sum_{i=0}^{i=f-1} \frac{\binom{F}{i}\binom{N-F}{n-i}}{\binom{N}{n}},$$

where N is the number of proteins in the network, F is the number of proteins in the network annotated with the function under consideration, n is the size of the cluster, and f is the number of proteins within the cluster annotated with that function.

Cluster analysis is of course widely used in many diverse applications. As a result, a large number of general clustering methods have been developed, and many of these have been applied to interactome data. Broadly speaking, we consider three types of methods for clustering interaction networks. The first group of methods cluster interaction networks using standard distance- or similarity-based clustering techniques; the key issue here is typically in deciding on a suitable distance or similarity measure between two proteins in an interaction network. The second group of methods are network-based hierarchical methods that attempt to partition the entire network. The third group of methods focus on identifying local clusters in the network. We also describe a number of approaches that cannot be easily classified in any of the previous groups.

We note that some methods use only local neighborhood information when clustering whereas others use more global features of the network; nevertheless, even when using local features to cluster proteins, clustering can be performed on the entire interactome, and thus in some sense, such clustering approaches incorporate the global organization of the interactome as well.

8.4.6.1 Distance-Based Clustering

In many attempts to cluster interactomes, distances or similarities between pairs of proteins are calculated, and then this distance or similarity matrix is used in conjunction with standard clustering approaches such as hierarchical clustering. Various similarity measures have been proposed for clustering interaction networks. In one approach [SL03], the similarity between two proteins is determined by considering each protein's interactions, and computing the significance of their number of shared interactions via a formula that is equivalent to the hypergeometric distribution. An alternate approach that also measures the overlap between the sets of interactions for each pair of proteins uses the Czekanowski-Dice distance [BCM+03]. For proteins u and v, this is given by:

$$CD(u, v) = \frac{|\mathcal{N}_1(u) \Delta \mathcal{N}_1(v)|}{|\mathcal{N}_1(u) \cup \mathcal{N}_1(v)| + |\mathcal{N}_1(u) \cap \mathcal{N}_1(v)|},$$

where Δ computes the symmetric difference between two sets. In addition to these two measures [SL03, BCM+03], there are several other ways of computing the similarity or distance between two proteins by considering only the overlap among their direct interactions [GR03, KvMB03]. In contrast to these local measures, a more global measure can be used where the distance between two proteins is calculated as the shortest path distance between them in the network [AMM05]. In a related earlier approach [RG03], each protein is associated with a vector that contains its shortest path distance to all other proteins in the network. A similarity between two proteins is obtained by computing the correlation coefficient between their corresponding shortest-path vectors. Since global and local similarity measures may be quite different, this global shortest-path based similarity has also been used in conjunction with a local connectivity coefficient measuring the common interactors of two proteins [PH04].

For these methods, agglomerative hierarchical clustering is then performed. In general, these methods build a hierarchy among proteins by progressively merging groups of proteins that are closest or most similar to each other. The neighbor-joining method [SN87] has also been used in the context of clustering interactomes [BCM+03]; it favors merging items that are close to each other but also considers distances from the remaining items. Note that hierarchical clustering methods do not automatically give the final partitioning of the network. In [BCM+03], the separation into clusters is performed using existing biological process annotations, whereby each cluster must have at least half of its proteins annotated by the same term. This function is then transferred to the other proteins in the cluster.

In some applications of hierarchical clustering, there can be a problem where distances among several items are identical. This is clearly the case when setting the distance between two proteins as their shortest path distance in the network. One solution to this problem is a two phase approach [AMM05]. In the first phase, hierarchical clustering is performed many times, and each time there is a "tie in proximity," a random pair is chosen to merge. Each clustering process is stopped according to a threshold that considers the distances between all proteins in a cluster. In the second phase, the fraction of solutions in which each protein pair is clustered together is itself used as a similarity measure for a final round of clustering.

8.4.6.2 Network-Based Hierarchical Clustering

Girvan and Newman [GN02] introduce a divisive hierarchical clustering procedure, based on edge betweenness. The betweenness for an edge is defined as the number of shortest paths between all pairs of vertices that run through that edge. It is expected that edges between modules have more shortest paths through them than those within modules, and therefore have higher betweenness values. Their procedure to partition the network successively deletes edges with highest betweenness values. The edge-betweenness clustering procedure has been applied to yeast and human interaction data [DDS05]. The Girvan-Newman algorithm has also been modified so that shortest paths are computed on weighted networks, and instead of counting the total number of shortest paths through an edge, the total number of "non-redundant" shortest paths through an edge are counted by considering paths that do not share an endpoint [CY06]. Note that edge weights in this case correspond to distances and not reliabilities or similarities. An alternate approach has modified the Girvan-Newman algorithm by iteratively deleting the edge with lowest edge clustering coefficient [RCC+04]. The edge clustering coefficient is a generalization of the usual clustering coefficient (see Section 8.4.6.3), and measures the number of triangles to which a given edge belongs, normalized by the number of triangles that might potentially include it. To handle better the case where the edge is found in no triangles, the edge clustering coefficient for edge (u, v) is defined as:

$$ECC(u, v) = \frac{z_{u,v} + 1}{\min\{|\mathcal{N}_1(u)| - 2, |\mathcal{N}_1(v)| - 2\}},$$

where $z_{u,v}$ gives the number of triangles that edge (u, v) participates in. Unlike the edge betweenness measure, the edge clustering coefficient is a local measure; however, in principle, this definition can be extended to handle higher order cycles as well.

Divisive methods do not necessarily specify how to get modules or clusters from the hierarchical grouping process. One working definition of a module is to consider a set of vertices $V' \subset V$ as a module if, for each of its vertices, the number of interactions it has within V' (its indegree) is greater than the number of interactions it has with vertices in $V - V'$ (its outdegree) [RCC+04]. This condition can be weakened so that a module only requires that the sum of the indegrees for the all vertices in the module be greater than the sum of their outdegrees. The partitioning of the network can now be performed so that an edge with highest edge betweenness or lowest edge clustering coefficient is only removed if it results in two modules [RCC+04]. A slightly modified definition considers a set V' a module if the ratio of the number of edges within V' to the number of edges from vertices in V' to vertices outside of this set is greater than one [LYC+07]; this is almost the same criterion as that for a weak module [RCC+04], except that edges within V' are not counted twice. This definition has been used to uncover modules in an agglomerative procedure, where singleton vertices are considered initially and edges are added back in, using the reverse Girvan-Newman ordering, only if the edge is not between two modules.

8.4.6.3 Local Clustering

There are a number of approaches that attempt to isolate highly connected or dense components within the larger protein interaction network. The density of a set of vertices may be defined in many ways. The density of a set of vertices V' is sometimes computed as the total number of edges among the vertices in V' divided by the total number of possible edges within V' (i.e., $\binom{|V'|}{2}$). Finding the densest subgraph of a particular size is NP-hard (say by reduction from clique), and thus a number of heuristic approaches have been developed. In [SM03], a Monte Carlo procedure is developed that attempts to find a set of k nodes with maximum density. A special case of density is the clustering coefficient. It is computed for a vertex v as the density of the neighbors of v (i.e., $\mathcal{N}_1(v)$ with v excluded). In [BH03], each vertex is weighted using a measure similar to its clustering coefficient, but that instead attempts to exclude the effects of low-degree vertices. Low degree vertices are frequent in protein interaction networks, and may artificially lower the clustering coefficients of highly connected vertices in dense regions of the network that also happen to be connected to several vertices of low degree. The clustering coefficient is instead computed over a k-core of the neighbors of each vertex, where k-cores are maximal subgraphs of degree $\geq k$. The vertex with highest weight seeds the search process, and clusters are greedily grown out from it, with vertices being included in the cluster if their weights are above a given threshold. Once no more vertices can be added, this process is repeated for the next highest weighted unseen vertex in the network. A greedy graph clustering approach is also taken by [AUASM+06]. Here, a cluster is grown so as to maintain the density of the cluster above a particular threshold, as well as to ensure that each vertex that is added to the cluster is connected to a sufficient number of vertices already in the cluster. The process is initialized by finding the vertex that has the largest number of common neighbors with its neighbors (i.e., takes part in the largest number of triangles).

Instead of starting with the vertex that has highest weight according to some measure, the clustering process can also be seeded with a group of proteins [Bad03, AKGR04, MRW+05]. In the context of protein function prediction, these seed proteins can consist of proteins that are known to share some function or that are known to be part of the same complex. In [Bad03], each interaction is labeled with a confidence value or reliability between 0 and 1, and a protein is added to the cluster if there is a path from any seed protein to it such that the product of the reliabilities of the edges in the path is greater than a preselected threshold; for each candidate protein, this corresponds to computing its shortest path to any seed protein when mapping each edge reliability to its negative logarithm. Note that this approach scores the membership of a protein to the initial seed set using the probability of its connection via the single-most probable path. In [AKGR04], random networks are utilized to compute the probability that protein u is a member of the same functional group as the seed set of proteins. This probability is estimated by the fraction of random networks in which a path exists from u to any protein in the seed set. Each random network is generated by considering every edge in the original network, and adding it into the network with probability proportional to its reliability in the original network. This approach thus attempts to compute the probability of a connection to the initial seed set via any path in the network. In [MRW+05], the weight of each edge in the network corresponds to the probability of a functional relationship given the observed evidence for the pair of proteins. The initial set of seed proteins is first expanded to include a fixed number n_1 of proteins that have largest total weight on their edges to the proteins in the seed set. Then, all proteins that are within radius two from one of the proteins in the seed set are considered. The set is further expanded to include a fixed number n_2 of proteins that have largest connections to the query set when requiring all indirect paths to pass through the proteins selected in the previous step, and taking the connection weight of each such path as a product of the edge probabilities. This approach is thus ranking each protein first by its expected number of direct interactions to the seed set, and then by its expected number of indirect interactions through the first set.

Spectral analysis has also been applied to find dense substructures within protein networks [BZC+03]. Here, eigenvalues and eigenvectors of the adjacency matrix of the network are computed. For each positive eigenvalue, its eigenvector is used to group together proteins. In particular, the proteins corresponding to the larger components of the eigenvector tend to form dense subgraphs. Groupings are further required to be of sufficient size and have large enough interconnectivity.

8.4.6.4 Other Clustering Approaches

In [KPJ04], an initial random partitioning is modified by iteratively moving one protein from one cluster to another in order to improve the clustering's cost. The cost measure considers for each protein the number of proteins within its assigned cluster with which it does not interact, as well as the number of interactions from it to proteins not assigned to its cluster; both should be small in ideal clusterings. In order to avoid local minima, this local search is augmented by occasionally dispersing the contents of a cluster at random. Additionally, a list of forbidden moves is kept to prevent cycling back to a previous partition.

In [APF+06], k-clique percolation clusters are found. A k-clique is a complete subgraph over k nodes, and two k-cliques are considered adjacent if they share exactly $k-1$ nodes. A k-clique percolation cluster consists of all nodes that can be reached via chains of adjacent k-cliques from each other. One advantage of such an approach is that each protein can belong to several clusters. Given that a protein can have different roles in the cell, membership in several clusters is biologically meaningful, and it may be useful to identify a strategy that can recover multiple functions.

A clustering method based on (modified) random walks within a network has also been developed [vD00, EDO02]. The protein interaction network is transformed into a Markov process, where transition probabilities from u to v and v to u are associated with each edge (u, v) in the network; that is, the adjacency matrix is converted to a stochastic matrix. The stochastic-flow algorithm alternates between an expansion step, which causes flow to dissipate within clusters, and an inflation step, which eliminates flow between different clusters. In the expansion step, the probability transition matrix is squared; this corresponds to taking another step in a random walk. In the inflation step, each entry in the stochastic matrix is raised to the r-th power and then normalized to ensure that the resulting matrix is stochastic again; for $r \geq 1$, the inflation step tends to favor higher probability transitions, and thus has the effect of boosting the probabilities of intra-cluster walks and demoting inter-cluster walks. This process continues until convergence, at which point the connected directed components are evident. Note that in this algorithm, it is the inflation step that differentiates it from the traditional way of taking random walks on a graph. This stochastic flow-based clustering procedure has been applied to a protein interaction network that has been transformed into a line graph [PLEO04]. Here, each vertex in the new graph represents an interaction in the original network, and any two vertices are adjacent if they share protein content (i.e., the corresponding interactions in original network involve a common protein). The line graph formulation allows the stochastic flow-based clustering to place each protein into several clusters.

8.5 Evaluation of Methods

A comprehensive comparative evaluation of how different network analysis methods perform for the task of function prediction has not been performed. We briefly outline a couple of previously proposed testing frameworks and showcase the performance of some of the reviewed methods in these frameworks. Overall, it is difficult to judge the comparative performance of different methods by surveying the literature. This is due in part to differences in the testing framework, the gold standard functional terms used, and the precise interaction networks under consideration.

We note that because of the availability and quality of interaction and functional annotation data, most of the existing testing has been performed in the baker's yeast *Saccharomyces cerevisiae*. Additionally, cross-validation testing is standard practice in all frameworks. That is, the annotations of one (or more) protein is treated as unknown, and the annotations of the remaining proteins, along with the network, are used to predict its annotations.

8.5.1 Testing Frameworks

In [DZM$^+$03], the following testing scheme is proposed. For each annotated protein u with at least one annotated interaction partner, it is assumed to be unannotated and its function is predicted. Let k_u be the number of known functions for protein u, p_u the number of predicted functions for protein u, and o_u the amount of overlap between the set of known and predicted functions. The precision (or positive predictive value) is defined as:

$$\text{Precision} = \frac{\sum_u o_u}{\sum_u p_u}.$$

The recall (or sensitivity) is defined as:

$$\text{Recall} = \frac{\sum_u o_u}{\sum_u k_u}.$$

In follow up work [DTSC04], 134 GO biological process terms are chosen for consideration based on whether they annotate more than 50 proteins and none of their child biological process terms annotate the same set of proteins. Since GO is a directed acyclic graph and functional terms can be related to each other via parent/child relationships, the authors suggest modifications to this basic scheme when working with functions in a hierarchy. A possible weakness in this framework is that proteins that have more annotations will have a larger effect on performance measurements.

In [NJA$^+$05], evaluation is performed per-protein. The MIPS [RZM$^+$04] functional hierarchy is used, with 72 biological process terms chosen from the second level of hierarchy. For each protein, if the top scoring function is above some threshold, it is the prediction for the protein. If a prediction is a known functional annotation, it is considered a true positive, and otherwise, it is a false positive. In the case of multiple top-scoring predictions, a protein's prediction is counted as a true positive if more than half of the predictions made for it are correct, and a false positive otherwise. This approach is taken as a compromise between two extreme cases. In the first case, a prediction for a protein can be counted as a true positive if at least one of the predictions made for it is correct; however, in this case, a method that predicts every protein to participate in every function would only have true positives in this framework. At the other extreme, a protein can be counted as a true positive if every prediction made for it is correct. This, however, would count as false positives those proteins that get many correct predictions and only one incorrect one. An alternate and perhaps better approach would be to compute the precision and recall per-protein, and then average the results over proteins.

In [BvH06], a number of clustering approaches are evaluated as to how well they recapitulate known yeast protein complexes. While this is not the same as assessing the performance of function prediction, there is likely to be some relationship between the two. The algorithms are run both on simulated networks where complexes are embedded into the graph, and edges are added and removed at various proportions, as well as on data sets obtained in high-throughput experiments. Performance is measured by computing recall values (i.e., for each complex, find the cluster which has the highest fraction of its proteins) and precision values (i.e., for each cluster, find the maximal fraction of its proteins found in the same annotated complex). In theory at least, it is also possible to use either of the above approaches

[DZM+03, NJA+05] to evaluate how well the enriched biological processes in each cluster predict protein function.

8.5.2 Performance of Methods

In [NJA+05], the majority, neighborhood, multiway-cut and flow formulations are tested in two-fold cross-validation on the yeast proteome using ROC (receiver operating characteristic) analysis. Perhaps surprisingly, it is found that majority outperforms the more complicated neighborhood and multiway-cut formulations. The functional flow method outperforms majority (and the other methods) when considering proteins that interact with fewer than three annotated proteins of the same function, and otherwise it performs similarly to majority. Further analysis shows that the neighborhood method's performance deteriorates when considering radii greater than two, that the majority, multiway-cut and functional flow method improve when incorporating interaction reliability, and that multiple solutions are necessary for the cut-based method in order to get higher confidence predictions. The multiway-cut formulation was previously found to outperform the majority method [VFMV03]. However, the measure of success used to judge performance there was the fraction of times the top prediction for each protein is correct, and the score of the top prediction was not considered. ROC analysis, as in [NJA+05], with a varying threshold gives a more complete picture of performance, particularly with respect to high-confidence predictions, and shows that majority outperforms the cut-based method over a large false positive range, but the cut method is able to make predictions when majority cannot. A subsequent paper [MWK06] also finds that a cut-based approach does not outperform a strictly local approach which predicts function based on the fraction (instead of number) of neighbors with a particular function. In their case, the cut-based approach considered is the pairwise mincut problem of [KML+04].

In [DZM+03] and [CSW06], the authors find in leave-one-out testing that the Markov random field approach [DZM+03] outperforms the majority [SUF00] and neighborhood approaches [HNO+01] on the yeast interactome. However, the Markov random field approach is closely related to the pairwise cut-based approach except that different favorability values are permitted for intra-function interactions, and a prior probability of the occurrence of each function is used [DTSC04]. Thus, we might expect it to perform similarly to the cut-based methods. The added generality of the Markov random field approach over the cut-based approach may potentially explain why the former performs better than majority whereas the latter does not; however, a weakness with the testing as performed in [DZM+03, CSW06] is that the rank of a function for a protein is used as its score for the majority and neighborhood methods. This means, for example, that for the majority method, it does not matter in this testing framework if the top-scoring function for a protein appears nine times or one time among its direct interactions—both are treated equivalently. It remains to be seen whether the Markov random field approach will outperform the local method when scores—not ranks—are considered.

Overall, it is clear that the highest confidence biological process predictions can be made when a protein is interacting with many proteins with known annotation. In this case, a majority scheme performs well, as do other methods. On the other hand, when a protein is known to interact with only unannotated proteins, local approaches such as majority cannot make any predictions, whereas the cut, flow, Markov random field and clustering methods can. More broadly, for proteins with few interactions or few interactions with proteins with known annotations, more global methods are necessary for functional predictions. These methods are thus likely to be especially useful for characterizing proteins in unusual or less-studied proteomes.

Clustering methods have largely not been evaluated with respect to function prediction. However, the study of [BvH06] finds that the stochastic flow-based clustering procedure [vD00] is robust to alterations in the simulated data and clearly outperforms the other methods tested [BH03, KPJ04, BWD96] in extracting complexes from high-throughput physical interaction datasets.

8.6 Conclusions

In this paper, we have reviewed a number of methods that have been developed for predicting protein function using protein interaction networks. There are many interesting avenues for future research.

Moving forward, it is clear that a comprehensive evaluation for function prediction via network analysis is necessary. It is likely that different methods perform well in different circumstances, and ideally an evaluation would bring to light which method should be used in which situation. In particular, it should be possible to relate topological features of the network to performance. For example, local methods may be expected to perform well on dense networks. Additionally, the performance of a method may depend on how well annotated the proteome is. Since the experimentally determined interactomes of various organisms differ with respect to their coverage, network density, and known annotations, such an evaluation will be vital.

Here, we have focused our attention on methods for analyzing physical interaction networks. However, though it is not always possible due to data availability, it is clear that it is advantageous to incorporate and integrate information from several sources. For example, in [NJA$^+$05], a performance improvement is observed for all methods tested—majority, cut-based, and flow-based—when incorporating information about synthetic lethal interactions into physical interactomes. In well-studied organisms, such data integration may result in networks with many proteins with high degree, and local methods may work very well; however, in many organisms, the resulting networks will remain sparse, and thus more global methods are likely to be more useful in making novel functional predictions.

Most of the methods reviewed here for function prediction are based on some variant of guilt-by-association. However, biological "cross-talk" is evident in interactomes [SUF00], as there are many pairs of different biological processes recurring as annotations for interacting proteins. Future methods for predicting protein function will benefit from leveraging these types of interplay relationships between biological processes. In particular, as focus shifts towards trying to predict more specific biological process terms for proteins, guilt-by-association will become less powerful, as proteins performing specific different tasks work together to perform some more general process. Similarly, the development of methods that more directly exploit the functional hierarchy is likely to be fruitful. Promising research along both of these lines has been initiated [KOY06, CP06].

To conclude, since large-scale physical interaction data have been available for less than a decade, the development of approaches for predicting protein function via analysis of interaction networks is just beginning, and there are many opportunities for future work. At the same time, the underlying graph-theoretic formulations of the problem allow us to build upon decades of algorithmic and methodological advances. With more sophisticated method development and careful testing, it is clear that these types of network analysis approaches will play a significant role in suggesting protein function and guiding future experimental verification.

Acknowledgments The author thanks Elena Nabieva and Jimin Song for helpful discussions and for comments on the manuscript. This work has been supported by NSF CCF-0542187, NSF IIS-0612231, NIH GM076275, and the NIH Center of Excellence grant P50 GM071508.

References

[ABB+00] M. Ashburner, C. Ball, J. Blake, D. Botstein, H. Butler, J. Cherry, et al. Gene ontology: tool for the unification of biology. The gene ontology consortium. *Nat. Genet.*, 25(1):25–29, 2000.

[AKGR04] S. Asthana, O. King, F. Gibbons, and F. Roth. Predicting protein complex membership using probabilistic network reliability. *Genome Res.*, 14:1170–1175, 2004.

[AMM05] V. Arnau, S. Mars, and I. Marin. Iterative cluster analysis of protein interaction data. *Bioinformatics*, 21:364–378, 2005.

[APF+06] B. Adamcsek, G. Palla, I. Farkas, I. Derenyi, and T. Vicsek. Cfinder: locating cliques and overlapping modules in biological networks. *Bioinformatics*, 22: 1021–1023, 2006.

[AS06] T. Aittokallio and B. Schwikowski. Graph-based methods for analysing networks in cell biology. *Briefings in Bioinformatics*, 7:243–255, 2006.

[AUASM+06] M. Altaf-Ul-Amin, Y. Shinbo, K. Mihara, K. Kurokawa, and S. Kanaya. Development and implementation of an algorithm for detection of protein complexes in large interaction networks. *BMC Bioinformatics*, 7:207, 2006.

[Bad03] J. Bader. Greedily building protein networks with confidence. *Bioinformatics*, 19:1869–1874, 2003.

[BCM+03] C. Brun, F. Chevenet, D. Martin, J. Wojcik, A. Guenoche, and B. Jacq. Functional classification of proteins for the prediction of cellular function from a protein-protein interaction network. *Genome Biol.*, 5:R6, 2003.

[BH03] G. Bader and C. Hogue. An automated method for finding molecular complexes in large protein interaction networks. *BMC Bioinformatics*, 4:2, 2003.

[BK03] A. Bauer and B. Kuster. Affinity purification-mass spectrometry. *Eur. J. Biochem.*, 270:570–578, 2003.

[Bur98] C. Burges. A tutorial on support vector machines for pattern recognition. *Data Mining and Knowledge Discovery*, 2(2):121–167, 1998.

[BvH06] S. Brohee and J. van Helden. Evaluation of clustering algorithms for protein-protein interaction networks. *BMC Bioinformatics*, 7:488, 2006.

[BWD96] M. Blatt, S. Wiseman, and E. Domany. Superparamagnetic clustering of data. *Phys. Rev. Lett.*, 76:3251–3254, 1996.

[BZC+03] D. Bu, Y. Zhao, L. Cai, H. Xue, X. Zhu, H. Lu, et al. Topological structure analysis of the protein-protein interaction network in budding yeast. *Nucl. Acids. Res.*, 31:2443–2450, 2003.

[CHRZ07] Y.-R. Cho, W. Hwang, M. Ramanathan, and Aidong Zhang. Semantic integration to identify overlapping functional modules in protein interaction networks. *BMC Bioinformatics*, 8:265, 2007.

[CLR90] Thomas H. Cormen, Charles E. Leiserson, and Ronald L. Rivest. *Introduction to Algorithms*. MIT Press/McGraw-Hill, 1990.

[CP06] S. Carroll and V. Pavlovic. Protein classification using probabilistic chain graphs and the Gene Ontology structure. *Bioinformatics*, 22:1871–1878, 2006.

[CSW06] H. Chua, W.-K. Sung, and L. Wong. Exploiting indirect neighbors and topological weight to predict protein function from protein-protein interactions. *Bioinformatics*, 22:1623–1630, 2006.

[CY06] J. Chen and B. Yuan. Detecting functional modules in the yeast protein-protein interaction network. *Bioinformatics*, 22:2283–2290, 2006.

[DCS03] M. Deng, T. Chen, and F. Sun. An integrated probabilistic model for functional prediction of proteins. In *Proc. 7th Annual RECOMB*, pages 95–103. ACM, 2003.

[DDS05] R. Dunn, F. Dudbridge, and C. Sanderson. The use of edge-betweenness clustering to investigate biological function in protein interaction networks. *BMC Bioinformatics*, 6:39, 2005.

[DJP+92] E. Dalhaus, D. S. Johnson, C. Papadimitriou, P. Seymour, and M. Yannakakis. The complexity of the multiway cuts. In *Proc. 24th Annual STOC*, pages 241–251. ACM, 1992.

[DSC03] M. Deng, F. Sun, and T. Chen. Assessment of the reliability of protein-protein interactions and protein function prediction. In *Pac. Symp. Biocomput.*, pages 140–151, 2003.

[DTSC04] M. Deng, Z. Tu, F. Sun, and T. Chen. Mapping gene ontology to proteins based on protein-protein interaction data. *Bioinformatics*, 20:895–902, 2004.

[DZM+03] M. Deng, K. Zhang, S. Mehta, T. Chen, and F. Sun. Prediction of protein function using protein-protein interaction data. *J. Computational Biol.*, 10:947–960, 2003.

[EDO02] A. Enright, S. Van Dongen, and C. Ouzounis. An efficient algorithm for large-scale detection of protein families. *Nucleic Acids Res*, 30:1575–1584, 2002.

[FGK02] R. Fourer, D. M. Gay, and B. W. Kernighan. *AMPL: A Modeling Language for Mathematical Programming*. Brooks/Cole Publishing Company, Pacific Grove, CA, 2002.

[FS89] S. Fields and O.-K. Song. A novel genetic system to detect protein-protein interactions. *Nature*, 340:245–246, 1989.

[GK00] M. Galperin and E. Koonin. Who's your neighbor? New computational approaches for functional genomics. *Nat. Biotechnol.*, 18:609–613, 2000.

[GN02] M. Girvan and M. Newman. Community structure in social and biological networks. *Proc. Natl. Acad. Sci. USA*, 99:7821–7826, 2002.

[GR03] D. Goldberg and F. Roth. Assessing experimentally derived interactions in a small world. *Proc. Natl. Acad. Sci. USA*, 100:4372–4376, 2003.

[HHLM99] L. Hartwell, J. Hopfield, S. Leibler, and A. Murray. From molecular to modular cell biology. *Nature*, 402:C47–52, 1999.

[HNO+01] H. Hishigaki, K. Nakai, T. Ono, A. Tanigami, and T. Takagi. Assessment of prediction accuracy of protein function from protein–protein interaction data. *Yeast*, 18:523–531, 2001.

[ILO00] ILOG CPLEX 7.1, 2000. http://www.ilog.com/products/cplex/.

[JCB+04] T. Joshi, Y. Chen, J. Becker, N. Alexandrov, and D. Xu. Genome-scale gene function prediction using multiple sources of high-throughput data in yeast. *OMICS*, 8:322–333, 2004.

[JYG+03] R. H. Jansen, H. Yu, D. Greenbaum, Y. Kluger, N. Krogan, S. Chung, et al. A Bayesian networks approach for predicting protein-protein interactions from genomic data. *Science*, 302:449–453, 2003.

[KL02] R. Kondor and J. Lafferty. Diffusion kernels on graphs and other discrete input spaces. In *Proc. Intl. Conf. on Machine Learning*, pages 315–322, 2002.

[KML+04] U. Karaoz, T. M. Murali, S. Levotsky, Y. Zheng, C. Ding, C. R. Cantor, and S. Kasif. Whole-genome annotation by using evidence integration in functional-linkage networks. *Proc. Natl. Acad. Sci. USA*, 101:2888–2893, 2004.

[KOY06] M. Kirac, G. Ozsoyoglu, and J. Yang. Annotating proteins by mining protein interaction networks. *Bioinformatics*, 22:e260–e270, 2006.

[KPJ04] A. King, N. Przulj, and I. Jurisica. Protein complex prediction via cost-based clustering. *Bioinformatics*, 20:3013–3020, 2004.

[KvMB03] R. Krause, C. von Mering, and P. Bork. A comprehensive set of protein complexes in yeast: mining large-scale protein-protein interaction screens. *Bioinformatics*, 19:1901–1908, 2003.

[LBC+04] G. Lanckriet, T. Bie, N. Cristianini, M. Jordan, and W. Noble. A statistical framework for genomic data fusion. *Bioinformatics*, 20:2626–2635, 2004.

[LDAM04] I. Lee, S. Date, A. Adai, and E. Marcotte. A probabilistic functional network of yeast genes. *Science*, 306(2):1555–1558, 2004.

[LK03] S. Letovsky and S. Kasif. Predicting protein function from protein/protein interaction data: a probabilistic approach. *Bioinformatics*, 19 Suppl 1:i197–i204, 2003.

[LYC+07] F. Luo, Y. Yang, C. Chen, R. Chang, J. Zhou, and R. Scheuermann. Modular organization of protein interaction networks. *Bioinformatics*, 23:207–214, 2007.

[MRW+05] C. Myers, D. Robson, A. Wible, M. Hibbs, C. Chiriac, C. Theesfeld, et al. Discovery of biological networks from diverse functional genomics data. *Genome Biol.*, 6:R114, 2005.

[MWK06] T. Murali, C.-J. Wu, and S. Kasif. The art of gene function prediction. *Nat. Biotechnol.*, 24:1474–1475, 2006.

[NJA+05] E. Nabieva, K. Jim, A. Agarwal, B. Chazelle, and M. Singh. Whole-proteome prediction of protein function via graph-theoretic analysis of interaction maps. *Bioinformatics*, 21 Suppl. 1:i302–i310, 2005.

[PH04] J. Poyatos and L. Hurst. How biologically relevant are interaction-based modules in protein networks? *Genome Biol.*, 5:R93, 2004.

[PLEO04] J. Pereira-Leal, A. Enright, and C. Ouzounis. Detection of functional modules from protein interaction networks. *Proteins*, 54:49–57, 2004.

[RCC+04] F. Radicchi, C. Castellano, F. Cecconi, V. Loreto, and D. Parisi. Defining and identifying communities in networks. *Proc. Natl. Acad. Sci. USA*, 101(2):2658–2663, 2004.

[RG03] A. Rives and T. Galitski. Modular organization of cellular networks. *Proc. Natl. Acad. Sci. USA*, 100(2):1128–1133, 2003.

[RZM+04] A. Ruepp, A. Zollner, D. Maier, K. Albermann, J. Hani, M. Mokrejs, et al. The FunCat, a functional annotation scheme for systematic classification of proteins from whole genomes. *Nucleic Acids Res.*, 32:5539–5545, 2004.

[SL03] M. Samanta and S. Liang. Predicting protein functions from redundancies in large-scale protein interaction networks. *Proc. Natl. Acad. Sci. USA.*, 100:12579–12583, 2003.

[SM03] V. Spirin and L. A. Mirny. Protein complexes and functional modules in molecular networks. *Proc. Natl. Acad. Sci. USA.*, 100:12123–12128, 2003.

[SN87] N. Saitou and M. Nei. The neighbor-joining method: a new method for reconstructing phylogenetic trees. *Mol. Biol. Evol.*, 4:406–425, 1987.

[SSM03] E. Sprinzak, S. Sattath, and H. Margalit. How reliable are experimental protein-protein interaction data? *J. Mol. Biol.*, 327(2):919–923, 2003.

[SUF00] B. Schwikowski, P. Uetz, and S. Fields. A network of protein-protein interactions in yeast. *Nat. Biotechnol.*, 18:1257–1261, 2000.

[SUS07] R. Sharan, I. Ulitsky, and R. Shamir. Network-based prediction of protein function. *Molecular Systems Biology*, 3:88, 2007.

[TDO+03] O. Troyanskaya, K. Dolinski, A. Owen, R. Altman, and D. Botstein. A Bayesian framework for combining heterogeneous data sources for gene function prediction (in S. cerevisiae). *Proc. Natl. Acad. Sci. USA*, 100:8348–8353, 2003.

[TN04] K. Tsuda and W. Noble. Learning kernels from biological networks by maximizing entropy. *Bioinformatics*, 20 Suppl. 1:i326–i333, 2004.

[Vap98] V Vapnik. *Statistical Learning Theory*. Wiley, 1998.

[vD00] S. van Dongen. *Graph clustering by flow simulation*. PhD thesis, University of Utrecht, 2000.

[VFMV03] A. Vazquez, A. Flammini, A. Maritan, and A. Vespignani. Global protein function prediction from protein-protein interaction networks. *Nat Biotechnol.*, 21:697–700, 2003.

[vMHJ+03] C. von Mering, M. Huynen, D. Jaeggi, S. Schmidt, P. Bork, and B. Snel. STRING: a database of predicted functional associations between proteins. *Nucleic Acids Res.*, 31:258–261, 2003.

[vMKS+02] C. von Mering, R. Krause, B. Snel, M. Cornell, S. Oliver, S. Fields, and P. Bork. Comparative assessment of large-scale data sets of protein-protein interactions. *Nature*, 417:399–403, 2002.

[YFW03] J. Yedidia, W. Freeman, and Y. Weiss. Understanding belief propagation and its generalizations. In *Exploring artificial intelligence in the new millennium*, pp. 239–269. Morgan Kaufmann Publishers Inc., San Francisco, CA, USA, 2003.

[ZGS07] X. Zhu, M. Gerstein, and M. Snyder. Getting connected: analysis and principles of biological networks. *Genes Dev*, 21:1010–1024, 2007.

Chapter 9
Cross-Species Analysis of Protein-protein Interaction Networks

Nir Yosef, Eytan Ruppin, and Roded Sharan

Abstract Data on protein-protein interactions are increasing exponentially. To date, large scale protein interaction networks are available for human and most model species. The arising challenge is to organize these networks into models of cellular machinery. As in other biological domains, a comparative approach provides a powerful basis for addressing this challenge. In this chapter we review the on-going effort for analyzing protein-protein interaction networks and signalling pathways across species to infer conserved protein modules and predict protein function and interaction.

9.1 Introduction

Recent technological advances enable the systematic characterization of protein-protein interaction (PPI) networks across multiple species. Procedures such as yeast two-hybrid [1] and protein co-immunoprecipitation [2] are routinely employed nowadays to generate large-scale protein interaction networks for human and most model species [3,4,5,6,7]. An arising challenge is to organize the accumulating network data into models of cellular machinery. As in other biological domains, a comparative approach provides a powerful basis for addressing this challenge, allowing to overcome the high level of noise characterizing PPI data [8].

In this chapter we review current algorithms for protein network alignment and querying, focusing on the problem of protein complex identification. Comparative approaches to protein complex detection can be roughly classified into supervised and unsupervised ones. In a supervised setting, one is given a protein complex of interest and has to identify similar instances of this complex in another network. This gives rise to a network query problem. In an unsupervised setting, one is given two or more networks and has to identify regions that are conserved across the

N. Yosef
School of Computer Science, Tel-Aviv University, Tel-Aviv 69978, Israel
e-mail: niryosef@post.tau.ac.il

A. Panchenko, T. Przytycka (eds.), *Protein-protein Interactions and Networks*,
DOI: 10.1007/978-1-84800-125-1_9,

networks. The assumption is that conserved regions are likely to be biologically significant. This gives rise to a network alignment problem. We start by presenting techniques for comparing (aligning) two networks. We then discuss their generalizations to multiple networks. Next, we present methods for querying subnetworks within a network. Finally, we discuss methods to evaluate the quality of the inferred modules and present a case study in which we compare the performance of current alignment methods.

9.2 Preliminaries

We provide some basic graph theoretic notation and definitions that will be used throughout the review. For background on graph theory the reader is referred to [9,10].

Let $G = (V, E)$ be a graph (or equivalently, a network) with a vertex set V and an edge set E. We denote $V(G) = V$ and $E(G) = E$. G is called *connected* if there is a path between every pair of vertices. The degree of a vertex u is the number of edges incident to u. The degree sequence of a graph G is a sorted list of the degrees of its vertices. For a set of vertices $V' \subseteq V$, an *induced subgraph* with respect to V' is a subgraph whose vertex set is V' and whose edge set is $E' = \{(u, v) \in E : u, v \in V'\}$. The *distance* $d(u, v)$ between a pair of vertices u, v is the length (in edges) of the shortest path between them. We define the distance of a vertex from itself to be 0. For a vertex v, its *neighborhood* is $N(v) = \{u \in V : (u, v) \in E\}$. A path in a graph is a sequence of vertices such that from each of its vertices there is an edge to the next vertex in the sequence. A simple path is a path with no repeated vertices.

9.3 Methods for Pairwise Network Alignment

The network alignment problem calls for identifying network regions that are conserved in their sequence and interaction patterns across two or more species. While the general problem is hard, generalizing subgraph isomorphism, heuristic methods have been devised to tackle it. In this section we review methods for comparing a pair of networks. We present both local methods that try to identify conserved subsets of the networks (corresponding to conserved protein complexes or pathways), and global methods that aim at finding a global (and ideally a one-to-one) correspondence between the proteins in the two networks. In the next section we give a brief overview of methods for multiple network alignment. Throughout this section we denote by G_0 and G_1 a pair of PPI networks of two species 0 and 1.

9.3.1 Alignment-Graph Based Methods

One heuristic approach for the pairwise network alignment problem creates a merged representation of the two networks being compared, called a *network*

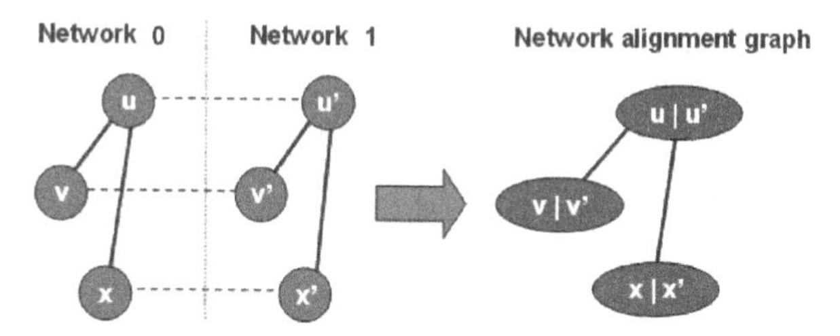

Fig. 9.1 Toy example of a network alignment. Networks 0 and 1 illustrate PPI networks of two species. Each *node* represents a protein, *solid lines* represent PPIs and *dotted horizontal lines* represent homology relationships between proteins from the two species. The alignment graph for the two networks appears on the *right*. Nodes represent pairs of sequence-similar proteins and edges represent conserved interactions

alignment graph, facilitating the search for conserved subnetworks. In a network alignment graph, the nodes represent pairs of proteins, one from each species, and the edges represent conserved PPIs across the two species, see example in Fig. 9.1. The alignment may consist of one-to-one correspondence between proteins across the two networks; however, in general there may be a many-to-many correspondence between proteins. This scenario can occur, for instance, when a single protein from one species is homologous to multiple proteins from the other species.

A network alignment graph provides the required framework for searching for conserved subnetworks, since these subnetworks will appear as subgraphs with specific structure in the graph. For instance, conserved protein complexes might appear as subgraphs of densely connected nodes. The heuristic was first used by Ogata et al. [11] when searching for correspondences between the reactions of specific metabolic pathways and the genomic locations of the genes encoding the enzymes catalyzing those reactions. Later on, Kelley et al. [12] applied this heuristic to study PPI networks. They translated the problem of finding conserved pathways to that of finding high-scoring paths in the alignment graph.

Alignment-graph based methods proceed in several phases. First, an alignment graph is constructed. Second, a procedure for scoring subgraphs of the alignment graph is defined. Last, a search heuristic is employed to identify high scoring subnetworks. In the rest of the section we will review three methods that use the alignment graph heuristic: NetworkBLAST [13] NetworkBLAST-E [14] and MaWish [15]. As these methods use similar graph construction and search procedures, we will mainly highlight their differences with respect to the scoring component.

9.3.1.1 The Network Alignment Graph

In a network alignment graph the nodes represent pairs of proteins, one from each species, and the edges represent conserved PPIs across the two species. The alignment between pairs of proteins from the two species is based on protein homology.

For every pair of homologous proteins, $u \in V(G_0)$ and $v \in V(G_1)$, a node $a = (u, v)$ is added to the alignment graph.

The definition of edges in the alignment graph somewhat varies between the different methods. In NetworkBLAST and NetworkBLAST-E, edges in the alignment graph represent *conserved interactions*, which are pairs of observed interactions, one in each species, between corresponding homologous proteins. More precisely, consider two nodes of the alignment graph, (u, u') and (v, v'), where $u, v \in V(G_0)$ and $u', v' \in V(G_1)$. The two nodes are linked if at least one of the pairs (u, v), (u', v') is observed to interact in its PPI network and the second spans proteins of distance at most two in the corresponding PPI network. See toy example of this data model in Fig. 9.1

The MaWish algorithm uses a wider definition according to which an edge linking two nodes (u, u') and (v, v'), can either represent a conserved interaction (as in Network BLAST), an interaction mismatch (where only one of those pairs directly interacts), or a duplication event (where at least one of the pairs (u, v), (u', v') represents paralogous proteins).

9.3.1.2 Search Heuristic

The alignment graph is used as a platform for the search of conserved protein complexes across multiple species. By construction, an induced subgraph C of the alignment graph corresponds to two species-specific sets of proteins C_0 and C_1, and can be assigned a score (or weight): $Score(C)$. A good scoring scheme should assign high scores to subnetworks that represent true conserved protein complexes.

The methods described in the following use different scoring schemes but conceptually similar heuristics to search for heavy (high scoring) subgraphs in the alignment graph. The search is performed in a bottom up manner, starting with small subgraphs as ''seeds". These seeds are then expanded by a greedy local procedure. For example, the search in the NetworkBLAST-E algorithm is performed as follows: For each node i in the alignment graph, identify a neighbor j such that the score of this pair is maximum among all neighbors of i. The algorithm enumerates all 4-node subgraphs which contain i and j and whose weight is above some threshold. These seeds are then greedily expanded, each time adding or deleting a node whose modification increases the weight of the current subgraph the most. The output of the search is a subnetwork C of the network alignment graph along with its corresponding score.

9.3.1.3 NetworkBLAST

This method considers the two subnetworks C_0 and C_1 as independent. Two models are defined, under which each of the subnetworks could have been created: a protein complex model, M_C, and a null model, M_N. Protein complexes are expected to be dense subnetworks, a property that is formulated in M_C by assuming that every edge appears with some high probability β independently of all other vertex pairs[1].

In the null model, M_N, it is assumed that the respective network (G_0 or G_1) was randomly selected from the collection of all networks with the same degree sequence. This induces a probability r_{uv} for each edge between two proteins (u, v). The value of r_{uv} is defined as the fraction of graphs with the same degree sequence as the original graph that contain an edge between u and v, and can be estimated analytically [16], or using a Monte Carlo approach [13]. Notably r_{uv} increases with the degrees of u and v, thus penalizing non-specific interactions between high degree nodes.

For a protein pair $(u, v) \in E_0 \cup E_1$, denote by T_{uv} the event that these two proteins interact, by F_{uv} the event that they do not interact, and by O_{uv} the set of observations on whether u and v interact. The probability that a given subnetwork C_i ($i \in \{0, 1\}$) was generated by the protein complex model (M_C) is:

$$P(C_i|M_C) = \prod_{(u,v)\in V_{C_i}} P(O_{uv}|M_C) = \prod_{(u,v)\in V_{C_i}} [\beta P(O_{uv}|T_{uv}) + (1-\beta)P(O_{uv}|F_{uv})]$$

Similarly, the probability that it was generated according to the null model (M_N) is:

$$P(C_i|M_N) = \prod_{(u,v)\in V_{C_i}} [r_{uv} P(O_{uv}|T_{uv}) + (1-r_{uv})P(O_{uv}|F_{uv})]$$

The estimation of $P(O_{uv}|T_{uv})$ relies on estimating the confidence we have in each putative interaction and the prior probability for an interaction [8].

Using these two models, the score of a subnetwork C is given by a log-likelihood ratio:

$$Score(C) = log\left(\frac{P(C_0|M_C)}{P(C_0|M_N)} \cdot \frac{P(C_1|M_C)}{P(C_1|M_N)}\right)$$

This formulation aims at distinguishing between a true significantly dense conserved protein complex and a random protein set. Thus, for example, it will give a relatively dense subnetwork that appears in a sparse area of the network a higher score than a dense subnetwork that appears in an area that is generally rich in interactions.

An example for a conserved complex identified between *S.cerevisiae* and *P.falciparum* [17] using NetworkBLAST is given in Fig. 9.2a.

[1] The parameter β can be estimated based on the density of known complexes. In practical applications of NetworkBLAST β is usually set to 0.8

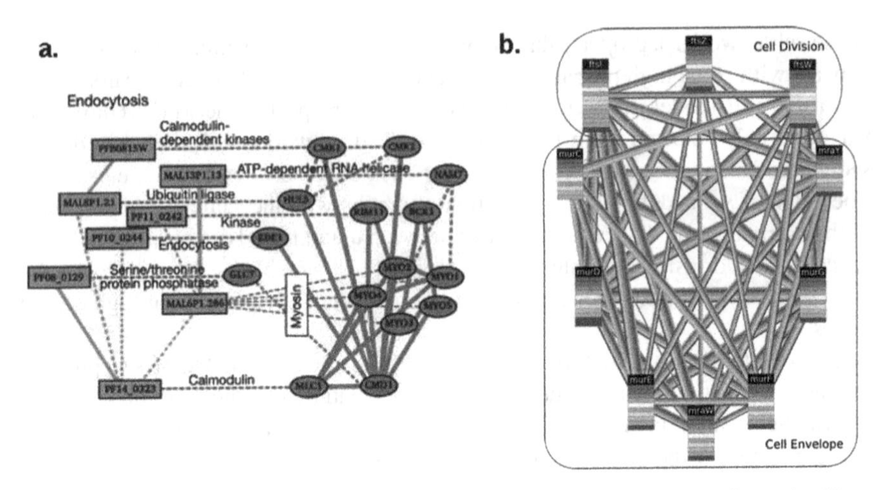

Fig. 9.2 Examples for pairwise and multiple network alignments. **(a)** A conserved complex identified between *S.cerevisiae* (*orange* nodes) and *P.falciparum* (*green* nodes). *Solid links* represent direct PPI, *dashed links* represent interactions mediated by one other protein. *Dashed horizontal lines* connect homologous proteins. The aligned complexes are annotated as components of endocytosis.This annotation was used by [17] to argue that the proteins PF10_0244 and MAL6P1.286 have a previously uncharacterized role in endocytosis. Reproduced with permission from Suthram et al. [17]. **(b)** A multiple network alignment including proteins from eight different bacterial networks. Each node is an equivalence class, labelled with its consensus gene name; *gray scale* bars within the nodes indicate the presence of the investigated species. Edges are colored using the same scheme as the nodes, and the width of each edge is proportional to its weight (for the list of investigated species and their color codes, see [20]). The list of proteins participating in this alignment is enriched with annotation of cell division and cell envelope. Reproduced with permission from Flannick et al. [20]

9.3.1.4 NetworkBLAST-E

NetworkBLAST-E is an extension of NetworkBLAST that aims at considering the evolutionary events that have led to the observed subnetworks, rather than scoring them independently. Two types of processes have been invoked to explain the evolution of PPI networks [18,19]: link dynamics and gene duplication. The first consists of sequence mutations in a gene that result in modifications of the interface between interacting proteins (Fig. 9.3a). Consequently, the corresponding protein may gain new connections or lose some of the existing connections to other proteins. The second consists of gene duplication, followed by either silencing of one of the duplicated genes or by functional divergence of the duplicates. The corresponding events in the network are the addition of a protein with the same set of interactions as the original protein, followed by the divergence of their links (Fig. 9.3b). Berg et al. [19] estimated the empirical rates of link dynamics and gene duplication in the yeast protein network, finding the former to be at least one order of magnitude higher than the latter. Based on this observation, they proposed a model for the evolution of protein networks in which link dynamics are the major evolutionary forces shaping the topology of the network, while slower gene duplication processes mainly affect its size.

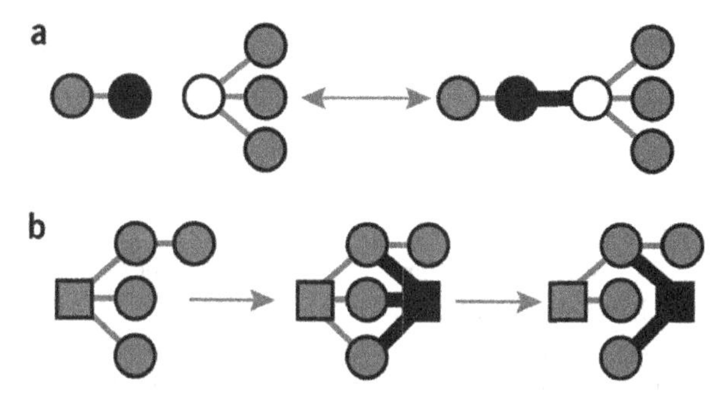

Fig. 9.3 Evolutionary processes shaping protein interaction networks. The progression of time is symbolized by arrows. **(a)** Link attachment and detachment occur through mutations in a gene encoding an existing protein. These processes affect the connectivity of the protein whose coding sequence undergoes mutation (shown in *black*) and of one of its binding partners (shown in *white*). **(b)** Gene duplication produces a new protein (*black square*) with initially identical binding partners (*gray square*). Empirical data suggest that duplications occur at a much lower rate than link attachment/detachment and that redundant links are lost subsequently (often in an asymmetric fashion), which affects the connectivities of the duplicate pair and of all its binding partners. Reproduced with permission from Berg et al. [19]

The difference between NetworkBLAST-E and NetworkBLAST is in the conserved protein complex model. The former assumes that the subnetworks C_0 and C_1 have evolved from a common ancestor through a series of duplication and link turnover events. Denote the common ancestor as S with a set of ancestral proteins V_S. The topology of S is assumed to be that of a dense subnetwork with edge probability β, as above. Let $\phi : V_{C_1} \cup V_{C_2} \rightarrow V_S$ be a mapping of the proteins to their ancestor in S. For each pair of proteins in the ancestral complex $(a, b) \in V_S$, let $L_{a,b}$ be the set of corresponding pairs in species 0, 1 under the mapping ϕ; namely $L_{a,b} = \{(u, v) \in V_0 : \phi(u) = a, \phi(v) = b\} \cup \{(u, v) \in V_1 : \phi(u) = a, \phi(v) = b\}$. Let D_i be the set of duplicated protein pairs (with the same ancestor according to ϕ) in species i. Finally, let P_L and P_G denote the probabilities of an interaction loss and gain events, respectively. The likelihood of the observed pair of subnetworks under this model is:

$$P(C_0, C_1|M_C) = \prod_{(a,b)\in V_S} P(O_{L_{ab}}|M_C) \prod_{(u,v)\in D_0 \cup D_1} P(O_{uv}|M_C)$$

where

$$P(O_{L_{ab}}|M_C) = \beta P(O_{L_{ab}}|T_{ab}) + (1-\beta)P(O_{L_{ab}}|F_{ab})$$

and

$$P(O_{L_{ab}}|T_{ab}) = \prod_{(u,v)\in Lab} P(O_{uv}|T_{uv})(1-P_L) + P(O_{uv}|F_{uv})P_L$$

$$P(O_{L_{ab}}|F_{ab}) = \prod_{(u,v)\in Lab} P(O_{uv}|T_{uv})P_G + P(O_{uv}|F_{uv})(1-P_G)$$

Thus, the probabilistic model of NetworkBLAST-E couples together the two subnetworks C_0 and C_1 and rewards conserved interactions.

9.3.1.5 MaWish

This method also uses an evolutionary based scoring scheme, which takes into account gene duplication and link turnover events. Every set of four proteins, two from each species ($u, v \in V_{G_0}$ and $u', v' \in V_{G_1}$) is given a weight $W(u, v, u', v')$, based on the probability that the proteins are true orthologs. Specifically, $W(u, v, u', v') = S(u, u') \cdot S(v, v')$, where $S(u, u') \in [0, 1]$ quantifies the likelihood that proteins u and u' are orthologous, and is computed based on their BLAST E-values.

When calculating the score for the given subnetwork C, two sets of quadruplets are defined; $M(C)$, contains all the quadruplets (u, v, u', v') where $u, v \in C_0$ and $u', v' \in C_1$ have a conserved interaction ($(u, v) \in E_0 \wedge d(u', v') \leq \Delta$ or $(u', v') \in E_1 \wedge d(u, v) \leq \Delta$) where Δ is a parameter of the algorithm, usually set to 1 or 2. $N(C)$, contains all the quadruplets for which an edge exists in one species and not conserved in the other. In addition, duplication events are treated as follows: let D_0 and D_1 be the sets of pairs of paralogous proteins in species 0 and 1, respectively. Every pair $(u, v) \in D_i$ is assigned a positive/negative duplication factor $d(u, v)$. Due to rapid functional divergence of duplicate proteins, in case the duplication occurred before the speciation event that split the two examined species, the authors wish to penalize it. Otherwise, in case it occurred after the speciation event, they wish to reward it. The authors employ sequence similarity as a means for distinguishing between events that occurred before and after the speciation event. This is based on the observation that sequence similarity provides a crude approximation for the age of duplication.

Finally, the score is calculated by summing over all quadruplets in $M(C)$ and $N(C)$. Conserved interactions increase the score by $\lambda \cdot W(u, v, u', v')$ and non-conserved interactions decrease the score by $\alpha \cdot W(u, v, u', v')$. Duplicate pairs of proteins $(u, v) \in D_0 \cup D_1$ reward/penalize the total weight according to $d(u, v)$. The score is formulated as follows:

$$
\begin{aligned}
Score(C) = & \sum_{(u,v,u',v') \in M(C)} \lambda \cdot W(u, v, u', v') - \\
& \sum_{(u,v,u',v') \in N(C)} \alpha \cdot W(u, v, u', v') + \\
& \sum_{(u,v) \in D_0} \mu \cdot d(u, v) + \sum_{(u',v') \in D_1} \mu \cdot d(u', v')
\end{aligned}
$$

where λ, α and μ are parameters of the algorithm (all have positive values).

9.3.2 Match-and-Split

The Match-and-Split method [21] uses a different approach from the alignment-graph based methods presented so far, and employs a top down search procedure to detect matching subsets of two given networks. By restricting its attention to a compact set of possible solutions (or pairs of matching subgraphs), the members of which are hypothesized to capture biologically plausible properties such as high connectivity, the method is able to enumerate all possible solutions in polynomial time. The compact set of solutions is defined as all subgraph pairs C_0, C_1 admitting the following properties:

Connectivity: C_0 and C_1 are connected.
Local Matching: Each node u in C_0 has at least one match v in C_1 and vice versa, where the match is decided according to a local criteria ($match_{C_0,C_1}(u, v) \in$ {true, false}) defined separately for each pair C_0, C_1.
Maximality: There does not exist any pair of subnetwork C'_0, C'_1, containing but not equal to C_0 and C_1 respectively, which satisfies the above two requirements.

A local matching function $match_{C_0,C_1}(\cdot, \cdot)$ is said to be *monotone* if all the matches it implies are preserved upon addition of more nodes to C_0 or C_1. By limiting the selection of allowable local matching functions to monotone ones, the authors gain an important advantage; combined with the maximality requirement, the monotonicity of the matching function implies a quadratic bound on the number of possible solutions (since any two solutions (C_0, C_1) and (C'_0, C'_1) with a common node pair $(u, v) \in (C_0 \cap C'_0, C_1 \cap C'_1)$ can be combined into a larger solution $(C_0 \cup C'_0, C_1 \cup C'_1)$).

Two locally matching proteins must have similar sequences and similar neighborhoods. One possible similarity definition is based on conserved paths of a fixed short length p (typically $p \leq 2$) which contain the two proteins. That is, $match_{C_0,C_1}(u, v)$ is true whenever some p-long path in C_0 containing u is similar to a p-long path in C_1 containing v; where two paths are considered similar if all pairs of corresponding proteins are sequence similar. It is easy to see that this similarity function is monotone. Other options for a local similarity measures are discussed in [21].

The search for pairs of matching subgraphs in G_0, G_1 is done by a recursive application of match and split operations:

Match: Remove all proteins in G_0 or G_1 with no match on the other network.
Split: Partition the remaining proteins into connected sets.

These operations are repeated for each pair of connected subgraphs (originated from G_0 and G_1, respectively) until they cannot be further splitted. The algorithm is guaranteed to run in polynomial time and enumerate all possible solutions.

In practice, the authors have noticed that some of the maximal solutions can get very large and, therefore, less plausible. To remedy this, these large solutions (which cannot be further splitted) are subject to an alternative split operation using some clustering technique different from the one normally used in the split stage.

9.3.3 Global Network Alignment

Unlike the above methods, which search for local highly matching regions between networks, methods for global network alignment search for the best overall match. Such an alignment is achieved by mapping the proteins in the two networks in a way that each protein is matched with at most one counterpart on the other network. The difference between the local and global approaches is analogous to the one between local and global sequence alignments. While the former allows us to detect conserved sequence motifs or network modules, the latter can be used to understand cross species variations on the level of the entire genome or interactome. An important property of the global network alignment problem is that it offers non-ambiguous solutions for the problem of pairing the proteins of one species with their functionally equivalent orthologs in another species. Each protein can be paired with a single best match, unlike in a local alignment solution where a single protein can be matched with several counterparts. One should note, however, that the non-ambiguity requirement might be too strict and that in certain cases, as in duplication after speciation, a many-to-many correspondence cannot be avoided.

The two global methods surveyed here rely on similar algorithmic ideas. Given two networks, an alignment graph is constructed and every node in the alignment graph is assigned with a score, reflecting the reliability of the implied orthology relation. These scores, in turn are computed based on graph-diffusion techniques, designed to capture global properties the alignment graph [22]. Intuitively, a pair of proteins gets a high score if their sequences are similar and if their neighbors on their respective PPI networks highly match to one another as well.

9.3.3.1 Orthology Detection Using Markov Random Fields

The Markov random field [23] model provides a probabilistic framework for simulating the mutual influence of random variables via a neighborhood system. Given a network of influence, the state of any random variable is assumed to be independent of all other random variable states given those of its immediate neighbors.

Bandyopadhyay et al. [24] used this method to model the probabilities for orthology relations between proteins in two different species. Each random variable i corresponds to a node in the network alignment graph and its state Z_i can be either "true", meaning that the corresponding proteins are true orthologs or "false", otherwise. The probability $P(Z_i)$ for the event that a given node i in the alignment graph represent a true orthology pair are conditioned with those events in its neighbors. This conditional probability is expressed as $P(Z_i|Z_{N(i)}) = \frac{1}{1+e^{-\alpha-\beta c(i)}}$, where the *conservation index* $c(i)$ reflects the ratio of true orthologies among the neighbors of node i in the alignment graph, and α and β are parameters that are learned in a supervised manner. The algorithm relies on the existence of a training set of known true and false orthologies represented in the alignment graph. With these constraints in hand, the posterior probabilities $P(Z_i)$ are estimated using the method of Gibbs sampling [25].

9.3.3.2 ISORank

The ISORank algorithm [26] assigns similarity scores to pairs of proteins from the two aligned networks according to a random walk model. Intuitively, a random walk in an alignment graph is simulated, and the score of a pair (i, j) reflects the probability of visiting its corresponding node. Formally, the score R_{ij} is calculated as follows:

$$R_{ij} = \alpha \sum_{u \in N(i)} \sum_{v \in N(j)} \frac{w(i, u)w(j, v)}{\sum_{r \in N(u), q \in N(v)} w(r, u)w(q, v)} R_{uv}$$

$$+ (1 - \alpha)E_{ij} \quad i \in V(G_0), \ j \in V(G_1)$$

where $w(u, v)$ is the reliability of the interaction (u, v), E_{ij} is a sequence similarity score, and $0 \leq \alpha \leq 1$ is a parameter which adjusts the relative effect of the local match E_{ij} on the overall score.

As mentioned above, the similarity score is derived from simulating a random walk in the alignment graph, where each node correspond to a pair of proteins (i, j), one from each species, and where edges connect two nodes (i, j), (i', j') if both (i, i') and (j', j') interact. At a given point during the walk, the probability to move to an adjacent node is proportional to the confidence of the respective interactions. In addition, with probability $1-\alpha$ the walk can be restarted from a randomly chosen node in the network, where the probability for a node to be selected in such a case is proportional to its respective sequence similarity score. The calculation of the scores R_{ij} is based on a reformulation as an eigenvalue problem in a similar manner to [27].

In addition to scoring all potential orthologous pairs, ISORank also computed an optimal global orthology mapping. This mapping, denoted Φ, is efficiently extracted from R in a way that each protein is matched with at most one orthologous counterpart and the sum $\sum_{(i,j)\in\Phi} R_{ij}$ is maximized. While their method was not explicitly designed to detect conserved protein clusters, the authors of [26] noted that the connected components in the global alignment network induced by Φ often correspond to conserved functional modules.

9.4 Multiple Network Alignment

The generalization of the network alignment process to more than two networks entails devising an appropriate scoring scheme and extending the notion of a network alignment graph. Stuart et al. [28] tackled the latter problem in the context of cross-species co-expression networks by forcing a consistent 1-1 mapping across all the networks, obtaining an alignment graph in which each gene is a member of at most one node. Another relatively simple scenario occurs when the compared networks are linear paths. The network alignment problem then becomes completely analogous to the sequence case, and one could adapt multiple sequence alignment techniques, such as progressive alignment, for its solution [29]. Recently, Sharan et al. [13] described a framework for multiple network alignment, which handles general correspondence relationships across networks. The scoring scheme extends the likelihood approach described above. The search problem is handled by extending the notion of a network alignment graph to multiple networks, albeit with an increased computational complexity, which scales as nh^{k-1} for k networks of size n with an average number of h possible orthologs to a protein per species. This method was applied to systematically identify conserved protein subnetworks across yeast, worm and fly, uncovering 71 conserved network regions that fell into well-defined functional categories.

9.4.1 Græmlin

The Græmlin algorithm [20] was designed to perform multiple network alignments as well as network querying tasks, and is guided by an evolutionary based scoring scheme. To perform multiple alignment, Græmlin uses an analog of the progressive sequence alignment technique. At each step, the algorithm aligns a pair of networks that are closest on the phylogenetic tree relating the analyzed species. These two networks are subsequently replaced by a collection of highly matching (conserved) subgraphs that resulted from their alignment. In each pairwise alignment step of this incremental procedure Græmlin searches for highly matching pairs of nodes, one of each network, and subsequently uses them as seeds in a greedy search procedure, similar to those discussed above.

Due to the incremental nature of the alignment procedure, the nodes in an input network might represent sets of homologous proteins (matched at a previous stage of the algorithm) rather than single proteins. Following the notation of the authors, we call these sets *equivalence classes*. An alignment of two subnetworks C_0, and C_1 therefore induces a new set of equivalence classes, obtained by unifying the sets of homologous proteins represented by the matched nodes.

The Græmlin's scoring framework is based on the sum of two components: node scores (assigned for each of the equivalence classes) and edge scores (assigned to edges between proteins of different classes). Both are based on the evolutionary interpretation of an alignment where each equivalence class represents a set of proteins which descended from a common ancestor. For each component of the scoring function, two probabilistic models are defined: an alignment model Ω according to which the aligned subnetworks were subject to evolutionary constraints, and a null model $\Re$ which assumes there were no such constraints.

The score assigned with a unified equivalence class u is based on the reconstruction of the evolutionary events that led from a single hypothesized ancestral protein to the proteins in the class. This score is computed as:

$$Score(u) = \sum_{x,y \in u} \alpha_{x,y} log \frac{Pr_{\Omega}^{mut}(x,y)}{Pr_{\Re}^{mut}(x,y)} + \sum_{event} log \frac{Pr_{\Omega}^{event}(n_u(event))}{Pr_{\Re}^{event}(n_u(event))}$$

The first term of the score concerns the event of sequence mutation. Pr_{Ω}^{mut} is a distribution of BLAST scores of orthologous pairs [30], while $Pr_{\Re}^{mut}$ is a distribution over random protein pairs. The score of each pair of proteins $x, y \in u$ is weighted by a factor $\alpha_{x,y}$, determined according to the evolutionary distance between their species (as in [31). The second term in the score is a sum over the likelihood ratios of four additional events: insertions, deletions, duplications and divergences. The number of occurrences of each event ($n_u(event)$) is approximated based on the most parsimonious evolutionary history of the unified equivalence class u.

The score of an edge between proteins of two unified equivalence classes i and j is computed as:

$$Score(e) = \frac{Pr_{\Omega}^{ij}(w(e))}{Pr_{\Re}^{ij}(w(e))}$$

where the weight of an edge $w(e)$ specifies the probability that the two corresponding proteins interact. The distribution functions of the alignment model are defined via an *edge scoring matrix* M. This is a symmetric matrix in which each row/column correspond to a different unified equivalence class, and where each cell M_{ij} specifies the corresponding $Pr_{\Omega}^{ij}(\cdot)$ probability distribution. The definition of M allows a user to specify the desired ancestral topology, be it a densely connected component, a path, or a specific topology determined by a given query subnetwork. A key stage prior to the computation of the edge score is the assignment of the equivalence classes with a specific row/column index in M. Ideally, one would choose an assignment which yields the maximum score. In practice, this assignment is approximated using a greedy search heuristic.

An example for a conserved complex identified from a set of ten bacterial networks [20] is given in Fig. 9.2b.

9.5 Network Querying

In contrast to network alignment, which aims at identifying significant subnetworks de-novo, network querying is a supervised task that aims at transferring knowledge from one network to another. The input to the problem is a known, typically well-researched subnetwork in one species, and the goal is to identify similar subnetworks in the network of another, typically less studied species. Similarity is measured both in terms of the protein sequences in the matched networks and in terms of the interaction patterns.

Most current algorithms define some sequence similarity threshold and seek an isomorphic or a homeomorphic match. In the former case the goal is to pair the query proteins with sequence-similar proteins from the network, so that the sub-graph induced by the paired vertices is isomorphic (identical) to the query. In the latter case, the match may contain unpaired vertices of degree 2 (and these are called insertions), and some of the query nodes of degree 2 may be unpaired as well (and these are called deletions).

Computationally, the querying problem is NP-hard as it generalizes the sub-graph isomorphism problem [32]. Hence, several restrictions of the problem have been studied, making it amenable to efficient solutions. These include restricting the structure of the network, restricting the structure of the query and relaxing the requirements of the sought matches.

Kelley et al. [12] were the first to address the query problem in the context of PPI networks. To tackle the problem they employed PathBLAST, a network alignment algorithm, by designating one of the aligned networks as the query. PathBLAST can handle queries that are linear paths with up to five proteins. It searches for conserved simple paths in the alignment graph, thereby identifying matching pathways. To allow flexibitility in the match, the alignment graph contains apart from direct edges (representing direct interactions in both species) also "gap" edges (representing direct interaction in one species and an indirect interaction through an intermediate protein in another species) and "mismatch" edges (representing indirect interactions in both species). Thus, the match implies the insertion of some proteins that do not match a query protein, and the deletion of some query proteins that are not matched. PathBLAST was applied successfully to query several known pathways in yeast.

In the following we review additional, more recent algorithms that are targeted toward the specific problem of network querying.

9.5.1 MetaPathwayHunter

MetaPathwayHunter [33] is a polynomial time algorithm for querying metabolic networks. It allows querying pathways that take the form of a multi source tree (a directed acyclic graph whose underlying undirected graph is a tree) in a target network that takes the form of a collection of multi-source trees.

Given a query network Q and a target network T the algorithm searches for subtrees of T that are homeomorphic to Q while allowing unmatched nodes only in the target network T. The different homeomorphisms between Q and T are scored according to the level of similarity of the nodes they are matching together and the number of implied insertion events. Consider a subtree $T' \subseteq T$ homeomorphic to the query Q and let $M[V(Q), V(T')]$ denote a homeomorphism-preserving mapping from the nodes of Q to the nodes of T' such that every node in Q is matched with exactly one node in T' and every node in T' is matched with at most one node from Q. The score of $M[V(Q), V(T')]$ is defined as:

$$\delta(|T'| - |Q|) + \sum_{(u,v)\in M} \Delta(u, v)$$

where δ is the insertion penalty and $\Delta(u, v)$ reflects the level of similarity between the proteins $u \in Q$, and $v \in T$. The high scoring homeomorphisms are efficiently found in a bottom-up manner, gradually expanding optimal alignments of subsets of the query and the target network using a dynamic programming procedure. With the assumption of Q and T being multi-source trees, this algorithm finds the optimal homeomorphic match in polynomial time. Handling inputs which do not conform to the topology restrictions of the algorithm is done by generating a set of multi-source trees that cover all the possible cycle splitting variations.

Given a query of a core pathway, the MetaPathwayHunter algorithm successfully revealed meaningful pathways in the target networks, e.g., an allantoin degradation pathway in E. Coli and an ureide degradation pathway in yeast. This work extends upon the linear pathways of PathBLAST to include tree-form queries, yet on the other hand, the target is not a general network but is rather assumed to be a forest of trees, or at least be easily recasted as one. Two other notable limitations of this method are that it ignores the variation in confidence levels of protein interactions, and does not account for deletion events.

9.5.2 QPath and QNet

QPath [34] is an algorithm for querying linear pathways developed by Shlomi et al. [34]. Unlike MetaPathwayHunter, in QPath no constraints are placed on the queried network. It aims at identifying matching paths while allowing for insertions of unmatched vertices from the network and deletions of query nodes.

The QPath algorithm relies on the color coding technique of Alon et al. [35] for ensuring that the discovered matches contain no vertex repetitions. Color coding is a randomized technique for efficient detection of simple, fixed length paths. For a given target network G and a query path of length k, the algorithm assigns a randomly chosen color from $\{1 \ldots k\}$ to every vertex in G, reducing the general problem of finding high scoring length-k paths to a simpler problem of finding high scoring paths that span distinct colors.[2] The latter problem can be solved in $2^{O(k)}m$ time, where m is the number of edges in G. This greatly improves upon the trivial $O(n^k)$ algorithm and is practical on current networks for $k \leq 12$.

Since any particular path may be assigned non-distinct colors and, hence, fail to be discovered, many random coloring trials are executed. The probability that a

---

[2]In practice, the number of colors is set to $k + N_{ins}$ where N_{ins} is the maximum number of allowed insertions.

given path is distinctly colored is $\frac{k!}{k^k} < e^{-k}$. Hence, if we execute $\ln(\frac{1}{\epsilon})e^k$ coloring iterations, we get the optimum path with probability at least $1 - \epsilon$ for any desired value of ϵ.

The matching pathways are scored according to their level of variation from the query pathway in terms of number of insertions and deletions, the sequence similarity of their constituent proteins to the query proteins, and the reliability of their constituent interactions. The weights of each of those factors in the overall score are estimated using logistic regression, where the regression aims at maximizing the fraction of matched pathways that are functionally enriched.

QPath was applied to query yeast pathways within the PPI network of fly. The resulting matches were found to be functionally coherent and, moreover, preserve the function of the corresponding queries, testifying to the utility of network querying in transferring biological annotations from one species to another.

Recently, Dost et al. [36] developed the QNet algorithm, which extends QPath to querying trees and tree-like structures. As in QPath, QNet searches for homeomorphic matches, and uses the color coding technique to prevent vertex repetitions within the matches. It relies on the notion of a tree decomposition, which intuitively is a representation of a given graph as a tree whose nodes represent subsets of vertices. A valid representation is one in which every edge of the graph occurs in one of the tree nodes (i.e., its endpoints are members of the corresponding subset), and the occurrences of every vertex span a subtree. The treewidth of the decomposition is the size of the largest subset minus one. For a query graph with treewidth t, the running time of QNet is $n^t 2^{O(k)}$.

Dost et al. further present a heuristic to query general graphs which is based on identifying many spanning trees of an input graph, querying each separately, and merging the results into a consensus match. This heuristic was successfully applied to query known protein complexes from yeast within the fly network.

9.5.3 PathMatch

A recent work by Yang et al. [37] reduced the problem of linear pathway querying in a general network to a substantially less complex problem – that of finding the longest path in a weighted directed acyclic graph. This simplification is facilitated by allowing the nodes in the target network to participate more than once in a solution. By leaving out the requirement for the solution to be a simple path, the PathMatch algorithm avoids the problem of cycles in the target network (which is the main complicating factor) and thus manages to operate in a polynomial time.

For a given query pathway $q = q_1 \ldots q_m$ and a target network $G = (V, E)$, denote by $V_i \subseteq V$ the set of proteins in V that may be associated with the query protein q_i. These correspondence lists are used to construct the directed acyclic graph G' on which the alignment search would be applied; The nodes in G' are partitioned into distinct subsets or *levels*. Level i in G' contains all the nodes from V_i (note that a node in G might appear more than once in G', on several different levels). The weight of a node u in level i is determined by its similarity to q_i. For each pair of nodes u, v from levels $i < j$ respectively, a directed edge (u, v) is placed when $j - i < m$ and $d_G(u, v) < m$ (where m is the maximum length of allowable insertions or deletions, and $d_G(\cdot, \cdot)$ is the distance function in the graph G). The weight of the edge is determined by the length of the insertion ($= d_G(u, v) - 1$) and deletion ($= j - i - 1$) that it implies.

From its construction it follows that the set of directed paths in G' covers all possible path alignments that are consistent with the correspondence lists V_i and do not contain a gap longer than m. A polynomial time algorithm is then used to efficiently find the highest scoring paths in G', where the score of an alignment is comprised of the similarity between the matched proteins (as reflected by the node weights), and the number of insertions and deletions (as reflected by the edge weights).

In contrast to PathMatch, the rest of the methods reviewed in this section do not allow multiple occurrences of a node in a solution (except pathBLAST where the simple path requirement is imposed only on the alignment graph which still makes it possible, though less likely, to have repeated nodes in a solution). The motivation for that is that very rarely does a protein take two different roles on the same pathway and hence protein repetitions are more likely to be biologically implausible.

An additional algorithm presented in [37] is GraphMatch – an exact algorithm for the network querying problem that does not place any restrictions on the topology of either the query or the target networks. GraphMatch operates in a brute force manner and enumerates all possible solutions (hence, its guaranteed accuracy). Naturally, this algorithm is highly complex and only applicable in very limited cases.

9.6 Evaluation Measures

9.6.1 Significance Evaluation

The common practice in most methods reviewed in this chapter, is to evaluate the statistical significance of the findings by comparison to randomized instances. The randomized instances, in turn, are generated by shuffling the edges of the participating interaction graphs while preserving vertex degrees, as well as shuffling the pairs of sequence-similar proteins while preserving the number of homologs per protein. Retaining these properties of the original data results in a more faithful random model.

A large number (typically ≥ 100) of randomized instances are analyzed and the resulting scores are then used to estimate a null distribution. The statistical significance of each solution found using the original data is evaluated against this null distribution, yielding empirical p-values which approximate the real probability to obtain such scores by chance.

The MaWish algorithm (see Section 9.3.1) employs a different technique to assign statistical significance to its findings. Here, a null model is defined based on the assumption that each interaction (within and across species) and each protein sequence is independent of all others. The mean and standard deviation of a score of a given alignment is then evaluated under the null model and compared to its original score (as entailed by the model used in the algorithm), producing a z-score that serves as an estimate for the significance of the alignment.

9.6.2 Quality Assessment

Several biologically-based measures for the goodness of a collection of protein modules have been suggested in the past. Here we review two types of widely used measures. The first quantifies the similarity between a given collection of protein modules and a reference, putatively true, catalog of protein complexes. The second type assesses the coherency of the conserved modules. It can be based on a number of sources such as the gene ontology (GO) annotation [38]) and gene expression profiles.

Importantly, both types of measures treat each species separately rather than explicitly evaluating the conservations hypothesis implied by each pair of aligned subnetworks. Such evaluation could, in principle, be made by comparing to a reference set of conserved modules. To date, however, most such references are not comprehensive enough and contain only a small number of cases to learn from. One exception is the Biocarta [39] database which contains many human-mouse conserved pathways.

1. *Similarity to a reference set*: To measure the level of correspondence with a reference set, we first need to define some measure for the match between individual complexes in the output collection and in the reference set. One way for example would be to evaluate the significance of a match using the following hypergeometric score:

$$Score(G,T,R) = \sum_{i=|V(R)\cap V(T)|}^{\min\{|V(R)|,|V(T)|\}} \frac{\binom{|V(R)|}{i}\binom{|V(G)|-|V(R)|}{V(T)-i}}{\binom{|V(G)|}{|V(T)|}}$$

where G is the entire protein network, T is the tested subnetwork, and R is the reference complex.

These significance levels are then corrected for multiple testing [40], and only cases which pass a desired false discovery rate threshold (typically 5%) are retained.

Now, denote by M the set of true complexes, and let $P = \cup_{m \in M} m$ be the set of proteins included in the complexes of M. Denote by H the collection of subnetworks to be examined such that $\forall h \in H, h \cap P \neq \emptyset$ and let $H^* \subseteq H$ be the subset of subnetworks that had a significant match in M. The *specificity* of the solution is defined as $|H^*|/|H|$. Let $M^* \subseteq M$ be the subset of complexes with a significant match in H. The *sensitivity* of the solution is defined as $|M^*|/|M|$.

A useful source of reference for these purposes is the MIPS catalog [41] of known yeast complexes (excluding category 550 which is obtained from high throughput experiments).
2. *Coherency of conserved clusters*: The coherence among proteins participating in the same subnetwork can be measured with respect to a number of properties and data sources. One property which we will later use to compare between some of the methods reviewed in this chapter, is the functional annotation of the protein set, as entailed by their GO functional annotation. A way to evaluate the enrichment of a GO term in a given set of proteins is via a hypergeometric score, as suggested in [13]. Such score should take into account ontology relations between terms. Specifically, since the GO terms are not independent but are rather connected by an ornithology of parent-child relationship, the enrichment of each term is conditioned on the enrichment of its parent term. The common practice is then to compare the enrichment scores to a null distribution of scores obtained with randomized data (in this case, random protein sets) and compute an empirical p-value.

9.7 A Case Study

In order to highlight the characteristics of each of the methods for pairwise alignment described above, we applied them to align two of the most established PPI networks: those of yeast and fly.

We downloaded protein interaction data for yeast and fly from the database of interacting proteins [42]. The yeast network contained 15,147 interactions spanning 4,738 proteins; the fly network contained 23,484 interactions spanning 7,165 proteins. We used a previously published logistic regression method [13] to assign reliabilities to the PPIs. The reliabilities were based only on the experimental evidence for each interaction.

We assessed the performance of the different methods by measuring the specificity, sensitivity and functional coherency of their suggested clusters. For comparison to a reference set, we downloaded the MIPS complex catalog (December 2005 download) and retained all complexes at level 3 or lower with at least one protein in the yeast PPI network (excluding category 550). Overall, there were 113 such complexes spanning 697 proteins. To assess the coherency of the identified clusters we extracted 4,818 and 6,140 GO *biological process* annotations for yeast and fly, respectively (December 2005 download).

Table 9.1 summarizes the performance of the four pairwise local network alignment methods reviewed above. It can be seen that NetworkBLAST-E and its predecessor NetworkBLAST perform quite similarly, and that both algorithms outperform the MaWish method both in terms of correspondence with the MIPS catalog and the coherence of the annotations of the proteins in the detected conserved clusters. Expectedly, the Match-and-Split algorithm has produced a substantially lower number of alignments due to its strict limitation to a compact set of solutions.

Table 9.1 Performance of the reviewed pairwise alignment methods with the yeast and fly PPI networks

				Functional enrichment	
Algorithm	#Complexes	Specificity (%)	Sensitivity (%)	Yeast (%)	Fly (%)
NetworkBLAST	146	74	19	79	46
NetworkBLAST-E	150	76	19	78	43
MaWish	97	69	13	67	38
Match-and-Split	24	91	8	91	75

These alignments, nevertheless, are highly accurate with high rates of specificity and coherence of annotation.

Recall that the major conceptual difference between NetworkBLAST and NetworkBLAST-E is that the former treats the aligned subnetworks independently from each other while the latter considers the evolutionary events which led to the emergence of the conserved subnetworks. Indeed, in [14] NetworkBLAST-E was shown to outperform NetworkBLAST when focusing on the conserved regions of the two networks being compared.

9.8 Discussion

The accumulation of protein interaction maps for multiple organisms makes network comparison a viable tool for predicting various properties of genes and proteins on a global scale. First and foremost, a conserved subnetwork that contains many proteins of the same known function suggests that the remaining proteins also have that function. Sharan et al. [13] used this concept to predict thousands of new protein functions for yeast, worm and fly, with an estimated success rate of 58–63%.

Network alignment is also instrumental in identifying functional orthologies; With the notion that similar protein sequences imply similar protein functions, the identification of orthologous pairs was traditionally applied based only sequence similarities [43]. One problem with this strategy is when the protein in question has similarity to not one but many paralogous proteins [44]. In these cases, every cross-species protein pair is technically orthologous but it is still necessary to distinguish which protein pairs play functionally equivalent roles [45]. Bandyopadhyay et al. [24] used the interactions made by the compared proteins (each in its own network) to determine the true orthology relations in these ambiguous cases. Network alignment was also successfully applied to infer other complex relationships, such as protein interactions or links between cellular processes [13,21].

Suthram et al. [17] used the pairwise network alignment methodology to measure the degree of evolutionary conservation between the pathogen *Plasmodium falciparum* and a number of eukaryotic model organisms by counting the number of significant alignments between their networks. Remarkably, the Plasmodium network had very few alignments with the yeast network and none with the rest of the inspected organisms, suggesting that this network encodes for many unique machineries.

A useful application of these findings is in guiding the design of new drugs, which will be better directed towards the unique characteristics of this parasite.

The application of network alignment techniques to network querying [33,34] has proven to be a very powerful tool for cross-species bootstrapping on the knowledge embedded in well-defined cellular networks, to identify parallel metabolic and signaling pathways in less researched networks.

There are a number of open challenges which lie ahead if one wishes to further develop and exploit the potential of alignment methods. For one, current algorithms for network querying are limited to sparse topologies such as paths and trees. Relaxing these limitations and handling more general queries is an important open problem. Other issues concerning alignment methods in general include their ability to handle alignments under partial, uncertain information and to be applied at lower computational costs. New efforts made towards this endeavor could benefit from the rich literature on graph theory [10], and on graph mining techniques in the data mining community [46,47]

Acknowledgments N.Y. was supported by the Tel-Aviv university rector and president scholarship. E.R. was supported by a MOST grant. R.S. was supported by an Alon Fellowship and by a research grant from the Israel Science Foundation (grant no. 385/06).

References

1. Ito T, Chiba T, and Yoshida M. Exploring the yeast protein interactome using comprehensive two-hybrid projects. *Trends Biotechnology*, 19:23–27, 2001.
2. Aebersold R and Mann M. Mass spectrometry-based proteomics. *Nature*, 422(6928):198–207, 2003.
3. Uetz P, Giot L, Cagney G, Mansfield TA, Judson RS, Knight JR, Lockshon D, Narayan V, Srinivasan M, Pochart P, et al. A comprehensive analysis of protein-protein interactions in Saccharomyces cerevisiae. *Nature*, 403(6770):623–627, 2000.
4. Ito T, Chiba T, Ozawa R, Yoshida M, Hattori M, and Sakaki Y. A comprehensive two-hybrid analysis to explore the yeast protein interactome. *Proc Natl Acad Sci USA*, 98:4569–4574, 2001.
5. Ho Y, Gruhler A, Heilbut A, Bader GD, Moore L, Adams SL, Millar A, Taylor P, Bennett K, Boutilier K, et al. Systematic identification of protein complexes in saccharomyces cerevisiae by mass spectrometry. *Nature*, 415(6868):180–183, 2002.
6. Gavin AC, Bosche M, Krause R, Grandi P, Marzioch M, Bauer A, Schultz J, Rick JM, Michon AM, Cruciat CM, et al. Functional organization of the yeast proteome by systematic analysis of protein complexes. *Nature*, 415(6868):141–147, 2002.
7. Stelzl U, Worm U, Lalowski M, Haenig C, Brembeck F H, Goehler H, Stroedicke M, Zenkner M, Schoenherr A, Koeppen S, et al. A human protein-protein interaction network: a resource for annotating the proteome. *Cell*, 122(6):957–968, 2005.
8. Deng M, Sun F, and Chen T. Assessment of the reliability of protein-protein interactions and protein function prediction. In Eighth Pacific Symposium on Biocomputing, p. 140–151, 2003.
9. Golumbic MC. *Algorithmic Graph Theory and Perfect Graphs*. Academic Press, New York, 1980.
10. Cormen TH, Leiserson CE, Rivest RL, and Stein C. *Introduction to Algorithms*. MIT Press and McGraw-Hill, 2001.

11. Ogata H, Fujibuchi W, Goto S, and Kanehisa M. A heuristic graph comparison algorithm and its application to detect functionally related enzyme clusters. *Nucleic Acids Res.*, 28(20): 4021–4028, 2000.
12. Kelley BP, Sharan R, Karp RM., Sittler T, Root DE, Stockwell BR, and Ideker T. Conserved pathways within bacteria and yeast as revealed by global protein network alignment. *Proc Natl Acad Sci*, 100(20):11394–11399, 2003.
13. Sharan R, Suthram S, Kelley RM, Kuhn T, McCuine S, Uetz P, Sittler T, Karp RM, and Ideker T. Conserved patterns of protein interaction in multiple species. *Proc Natl Acad Sci.*, 102(6):1974–1979, 2005.
14. Hirsh E and Sharan R. Identification of conserved protein complexes based on a model of protein network evolution. *Bioinformatics*, 23(2):e170–e176, 2007.
15. Koyuturk M, Grama A, and Szpankowski W. Pairwise local alignment of protein interaction networks guided by models of evolution. *J Comput Biol*, 13:182–199, 2006.
16. Itzkovitz S, Milo R, Kashtan N, Ziv G, and Alon U. Subgraphs in random networks. *Physical review E*, 68, 2003.
17. Suthram S, Sittler T, and Ideker T. The plasmodium protein network diverges from those of other eukaryotes. *Nature*, 438(7064):108–12, 2005.
18. Wagner A. The yeast protein interaction network evolves rapidly and contains few redundant duplicate genes. *Mol Biol Evol*, 18(7):1283–1292, 2001.
19. Berg J, Lassig M, and Wagner A. Structure and evolution of protein interaction networks: A statistical model for link dynamics and gene duplications. *BMC Evol Biol*, 4(1):51, 2004.
20. Flannick J, Novak A, Srinivasan BS, McAdams HH, and Batzoglou S. Græmlin: general and robust alignment of multiple large interaction networks. *Genome Research*, 16(9):1169–1181, 2006.
21. Narayanan M and Karp MR. Comparing protein interaction networks via a graph match-and-split algorithm. *J Comput Biol*, 14(7):892–907, 2007.
22. Weston J, Elisseeff A, Zhou D, Leslie CS, and Noble WS. Protein ranking: from local to global structure in the protein similarity network. *Proc Natl Acad Sci. USA*, 101:6559–6563, 2004.
23. Besag J. Spatial interaction and the statistical analysis of lattice systems. *J Roy Statist Soc*, B 55:192–236, 1993.
24. Bandyopadhyay S, Sharan R, and Ideker T. Systematic identification of functional orthologs based on protein network comparison. *Genome Res*, 16:426–35, 2006.
25. Smith A and Roberts G. Bayesian computation via the gibbs sampler and related markov chain monte carlo methods. *J Roy Statist Soc*, B 55:3–23, 1993.
26. Singh R, Xu J, and Berger B. Pairwise global alignment of protein interaction networks by matching neighborhood topology. *In The Proceedings of the 11th International Conference on Research in Computational Molecular Biology (RECOMB)*, pp. 16–31, 2007.
27. Page L and Brin S. The anatomy of a large scale hypertextual web search engine. *In Proceedings of the Seventh International World Wide Web Conference*, 1998.
28. Stuart JM, Segal E, Koller D, and Kim SK. A gene-coexpression network for global discovery of conserved genetic modules. *Science*, 302:249–255, 2003.
29. Tohsato Y, Matsuda H, and Hashimoto A. A multiple alignment algorithm for metabolic pathway analysis using enzyme hierarchy. *Proc Int Conf Intell Syst Mol Biol*, 8:376–183, 2000.
30. Tatusov RL, Koonin EV, and Lipman DJ. A genomic perspective on protein families. *Science*, 278(5338):631–637, 1997.
31. Altschul SF, Carroll RJ, and Lipman DJ. Weights for data related by a tree. *J Mol Biol*, 207(4):647–53, 1989.
32. Garey MR and Johnson DS. *Computers and Intractability: A Guide to the Theory of NP-Completeness.* W.H. Freeman and Co., 1979.
33. Pinter RY, Rokhlenko O, Yeger-Lotem E, and Ziv-Ukelson M. Alignment of metabolic pathways. *Bioinformatics*, 21:3401–3408, 2005.

34. Shlomi T, Segal D, Ruppin E, and Sharan R. QPath: a method for querying pathways in a protein-protein interaction network. *BMC Bioinformatics*, 7:199, 2006.
35. Alon N, Yuster R, and Zwick U. Color-coding. *J. ACM*, 42:844–856, 1995.
36. Dost B, Shlomi T, Gupta N, Ruppin, Bafna V, and Sharan R. Qnet: a tool for querying protein interaction networks. *In The Proceedings of the 11th International Conference on Research in Computational Molecular Biology (RECOMB)*, 2007.
37. Yang Q and Sze SH. Path matching and graph matching in biological networks. *Journal of Computational Biology*, 14(1):56–67, 2007.
38. Ashburner M, Ball CA, Blake JA, Botstein D, Butler H, Cherry JM, Davis AP, Dolinski K, Dwight SS, Eppig JT, et al. Gene ontology: tool for the unification of biology. the gene ontology consortium. *Nat Genet*, 25(1):25–29, 2000.
39. The Biocarta data base. http://www.biocarta.com/genes/allPathways.asp.
40. Benjamini Y and Hochberg Y. Controlling the false discovery rate: a practical and powerful approach to multiple testing. *J R Stat Soc*, 57(1):289–300, 1995.
41. Mewes HW, Amid C, Arnold R, Frishman D, Guldener U, Mannhaupt G, Munsterkotter M, Pagel P, Strack N, Stumpflen V, et al. MIPS: analysis and annotation of proteins from whole genomes. *Nucleic Acids Res*, 32 Database issue:D41–4, 2004.
42. Xenarios I, Salw'inski L, Joyce X, Higney P, Kim S, and Eisenberg D. Dip, the database of interacting proteins: a research tool for studying cellular networks of protein interactions. *Nucleic Acids Res*, 30(1):303–5, 2002.
43. Brenner SE. Errors in genome annotation. *Trends Genet*, 15:132–133, 1999.
44. Sjolander K. Phylogenomic inference of protein molecular function: Advances and challenges. *Bioinformatics*, 20(2):170–179, 2004.
45. Remm M, Storm CE, and Sonnhammer EL. Automatic clustering of orthologs and in-paralogs from pairwise species comparisons. *J Mol Biol*, 314:1041–1052, 2001.
46. Giugno R and Shasha D. Graphgrep: a fast and universal method for querying graphs. *In Proceeding of the International Conference in Pattern recognition (ICPR)*, 2002.
47. Koyuturk M, Grama A, and Szpankowski W. An efficient algorithm for detecting frequent subgraphs in biological networks. *Bioinformatics*, 20 (Supp 1):I200–207, 2004.

Index